Advanced Courses in Mathematics
CRM Barcelona

Centre de Recerca Matemàtica

Managing Editor:
Manuel Castellet

Ralph L. Cohen
Kathryn Hess
Alexander A. Voronov

String Topology and Cyclic Homology

Birkhäuser Verlag
Basel · Boston · Berlin

Authors:

Ralph L. Cohen
Department of Mathematics
Stanford University
Stanford, CA 94305-2125, USA
e-mail: ralph@math.stanford.edu

Alexander A. Voronov
School of Mathematics
University of Minnesota
Minneapolis, MN 55455, USA
e-mail: voronov@math.umn.edu

Kathryn Hess
Institut de Mathématiques
Faculté des Sciences de base EPFL
1015 Lausanne, Switzerland
e-mail: kathryn.hess@epfl.ch

2000 Mathematical Subject Classification: Primary: 57R19; 55P35; 57R56; 57R58; 55P25; 18D50; 55P48; 58D15; Secondary: 55P35; 18G55, 19D55, 55N91, 55P42, 55U10, 68P25, 68P30

A CIP catalogue record for this book is available from the Library of Congress, Washington D.C., USA

Bibliografische Information Der Deutschen Bibliothek
Die Deutsche Bibliothek verzeichnet diese Publikation in der Deutschen Nationalbibliografie; detaillierte bibliografische Daten sind im Internet über <http://dnb.ddb.de> abrufbar.

ISBN 3-7643-2182-2 Birkhäuser Verlag, Basel – Boston – Berlin

Printed on acid-free paper produced from chlorine-free pulp. TCF ∞
Printed in Germany
ISBN-10: 3-7643-2182-2
ISBN-13: 978-3-7643-2182-6
e-ISBN: 3-7643-7388-1

9 8 7 6 5 4 3 2 1

www.birkhauser.ch

Contents

Foreword

Free loop spaces play a central rôle in two recent advances in algebraic topology. The first one is string topology, a subject born with the seminal work of Chas and Sullivan in 1999, who uncovered new algebraic structure in the homology of free loop spaces on manifolds. The second one is topological cyclic homology, a topological version of Connes' cyclic homology introduced in 1993 by Bökstedt, Hsiang, and Madsen.

A summer school was held in Almería from September 16 to 20, 2003, to cover topics in this new and exciting field.

The first part of this book consists of the joint account of the two lecture series which focused on string topology (Cohen and Voronov). It discusses the loop product from the original point of view of Chas and Sullivan, from the Cohen-Jones stable point of view, as well as Voronov's operadic point of view.

The second part is essentially an account of the course devoted to the construction of algebraic models for computing topological cyclic homology (Hess). Starting with the study of free loop spaces and their algebraic models, it continues with homotopy orbit spaces of circle actions, and culminates in the Hess-Rognes construction of a model for computing spectrum cohomology of topological cyclic homology.

Ralph L. Cohen, Kathryn Hess, and Alexander A. Voronov

The summer school was made possible thanks to the support of the Groupement de Recherche Européen "Topologie Algébrique" (G.D.R.E. C.N.R.S. 1110) and the Centre de Recerca Matemàtica (Barcelona).

Part I

Notes on String Topology

Ralph L. Cohen and Alexander A. Voronov

Partially supported by a grant from the NSF

Introduction

String topology is the study of algebraic and differential topological properties of spaces of paths and loops in manifolds. It was initiated by the beautiful paper of Chas and Sullivan [CS99] in which algebraic structures in both the nonequivariant and equivariant homology (and indeed chains) of the (free) loop space, LM, of a closed, oriented manifold were uncovered. This has lead to considerable work by many authors over the past five years. The goals of this paper are twofold. First, this paper is meant to be an introduction to this new and exciting field. Second, we will attempt to give a "status report". That is, we will describe what has been learned over the last few years, and also give our views about future directions of research. This paper is a joint account of each of the author's lecture series given at the 2003 Summer School on String Topology and Hochschild Homology, in Almería, Spain.

In our view there are two basic reasons for the excitement about the development of string topology. First, it uses most of the modern techniques of algebraic topology, and relates them to several other areas of mathematics. For example, the description of the structure involved in the string topology operations uses such concepts as operads, PROPs, field theories, and Gerstenhaber and Batalin-Vilkovisky algebras. The fundamental role played by moduli spaces of Riemann surfaces in string topology, relates it to basic objects of study in algebraic and symplectic geometry. Techniques in low dimensional topology such as the use of graphs to study these moduli spaces are also used in an essential way. Moreover there are both formal and computational relationships between string topology and Gromov-Witten theory that are only beginning to be uncovered. Gromov-Witten theory is a basic tool in string theory, algebraic geometry, and symplectic geometry, and understanding its relationship to string topology is an exciting area of current and probably future research.

The second reason for the attention the development of string topology has been receiving has to do with the historical significance, in both mathematics and physics, played by spaces of paths and loops in manifolds. The systematic study of the differential topology of path and loop spaces began in the 1930's with Morse, who used his newly developed theory of "calculus of variations in the large" to prove among other things that for any Riemannian metric on the

n-sphere, there are an infinite number of geodesics connecting any two points. In the 1950's, R. Bott studied Morse theory on the loop spaces of Lie groups and symmetric spaces to prove his celebrated periodicity theorem. In the 1970's and 1980's, the K-theoretic tools developed by Waldhausen to study diffeomorphisms of high dimensional manifolds were found to be closely related to the equivariant stable homotopy type of the free loop space. Finally, within the development of string theory in physics, the basic configuration spaces are spaces of paths and loops in a manifold. Some of the topological issues this theory has raised are the following.

1. What mathematical structure should the appropriate notions of field and field strength have in this theory? This has been addressed by the notion of a "B-field", or a "gerbe with connection". These are structures on principal bundles over the loop space.
2. How does one view elliptic operators, such as the Dirac operator, on the loop space of a manifold? The corresponding index theory has been developed in the context of elliptic cohomology theory.
3. How does one understand geometrically and topologically, intersection theory in the infinite dimensional loop space and path space of a manifold?

It is this last question that is the subject of string topology. The goal of these notes is to give an introduction to the exciting developments in this new theory. They are organized as follows. In Chapter 1 we review basic intersection theory, including the Thom-Pontrjagin construction, for compact manifolds. We then develop and review the results and constructions of Chas and Sullivan's original paper. In Chapter 2 we review the concepts of operads and PROPS, discuss many examples, and study in detail the important example of the "cacti operad", which plays a central role in string topology. In Chapter 3 we discuss field theories in general, and the field theoretic properties of string topology. Included are discussions of "fat graphs", and how they give a model for the moduli space of Riemann surfaces, and of "open-closed" string topology, which involves spaces of paths in a manifold with prescribed boundary conditions. In Chapter 4 we discuss a Morse theoretic interpretation of string topology, incorporating the classical energy functional on the loop space, originally studied by Morse himself. In this chapter we also discuss how this perspective suggests a potentially deep relationship with the Gromov-Witten theory of the cotangent bundle. Finally in Chapter 5 we study similar structures on spaces of maps of higher dimensional spheres to manifolds.

Acknowledgments . We are very grateful to David Chataur, José Luis Rodríguez, and Jérôme Scherer for organizing and inviting us to participate in such an active and inspiring summer school. We would also like to thank David Chataur, Jérôme Scherer, and Jim Stasheff for many helpful suggestions regarding an earlier draft of the manuscript.

Chapter 1

Intersection theory in loop spaces

String topology is ultimately about the differential and algebraic topology of spaces of paths and loops in compact, oriented manifolds. The basic spaces of paths that we consider are $C^\infty(\mathbb{R}, M)$, $C^\infty([0,1], M)$, which we denote by $\mathcal{P}(M)$, $C^\infty(S^1, M)$, which we denote by LM, and $\Omega(M, x_0) = \{\alpha \in LM : \alpha(0) = x_0\}$. By the C^∞ notation we actually mean spaces of *piecewise* smooth maps. For example a map $f : [x_0, x_k] \to M$ is *piecewise smooth* if f is continuous and if there exists $x_0 < x_1 < \cdots < x_{k-1} < x_k$ with $f_{|(x_i, x_{i+1})}$ infinitely differentiable for all i. These spaces of paths are infinite dimensional smooth manifolds. See, for example [Kli82].

The most basic algebraic topological property of closed, oriented manifolds is Poincaré duality. This manifests itself in a homological intersection theory. In their seminal paper, [CS99], Chas and Sullivan showed that certain intersection constructions also exist in the chains and homology of loop spaces of closed, oriented manifolds. This endows the homology of the loop space with a rich structure that goes under the heading of "string topology".

In this chapter we review Chas and Sullivan's constructions, as well as certain homotopy theoretic interpretations and generalizations found in [CJ02], [CG04]. In particular we recall from [Coh04b] the ring spectrum structure in the Atiyah dual of a closed manifold, which realizes the intersection pairing in homology, and recall from [CJ02] the existence of a related ring spectrum realizing the Chas-Sullivan intersection product ("loop product") in the homology of a loop space. We also discuss the relationship with Hochschild cohomology proved in [CJ02], and studied further by [Mer03], [FMT02], as well as the homotopy invariance properties proved in [CKS05]. We begin by recalling some basic facts about intersection theory in finite dimensions.

1.1 Intersections in compact manifolds

Let $e : P^p \subset M^d$ be an embedding of closed, oriented manifolds of dimensions p and n respectively. Let k be the codimension, $k = d - p$.

Suppose $\theta \in H_q(M^d)$ is represented by an oriented manifold, $f : Q^q \to M^d$. That is, $\theta = f_*([Q])$, where $[Q] \in H_q(Q)$ is the fundamental class. We may assume that the map f is transverse to the submanifold $P \subset M$, otherwise we perturb f within its homotopy class to achieve transversality. We then consider the "pull-back" manifold

$$Q \cap P = \{x \in Q : f(x) \in P \subset M\}.$$

This is a dimension $q - k$ manifold, and the map f restricts to give a map $f : Q \cap P \to P$. One therefore has the induced homology class,

$$e_!(\theta) = f_*([Q \cap P]) \in H_{q-k}(P).$$

More generally, on the chain level, the idea is to take a q-cycle in M, which is transverse to P in an appropriate sense, and take the intersection to produce a $q - k$-cycle in P. Homologically, one can make this rigorous by using Poincaré duality, to define the intersection or "umkehr" map,

$$e_! : H_q(M) \to H_{q-k}(P)$$

by the composition

$$e_! : H_q(M) \cong H^{d-q}(M) \xrightarrow{e^*} H^{d-q}(P) \cong H_{q-k}(P)$$

where the first and last isomorphisms are given by Poincaré duality.

Perhaps the most important example is the diagonal embedding,

$$\Delta : M \to M \times M.$$

If we take field coefficients, the induced umkehr map is the *intersection pairing*

$$\mu = \Delta_! : H_p(M) \otimes H_q(M) \to H_{p+q-d}(M).$$

Since the diagonal map induces cup product in cohomology, the following diagram commutes:

$$\begin{array}{ccc} H_p(M) \otimes H_q(M) & \xrightarrow{\mu} & H_{p+q-d}(M) \\ {\scriptstyle P.D}\downarrow & & \downarrow{\scriptstyle P.D} \\ H^{d-p}(M) \times H^{d-q}(M) & \xrightarrow{cup} & H^{2d-p-q}(M) \end{array}$$

In order to deal with the shift in grading, we let $\mathbb{H}_*(M) = H_{*+d}(M)$. So $\mathbb{H}_*(M)$ is nonpositively graded.

Proposition 1.1.1. *Let k be a field, and M^d a closed, oriented, connected manifold. Then $\mathbb{H}_*(M^d; k)$ is an associative, commutative graded algebra over k, together with a map $\epsilon : \mathbb{H}_*(M; k) \to k$ such that the composition*

$$\mathbb{H}_*(M) \times \mathbb{H}_*(M) \xrightarrow{\mu} \mathbb{H}_*(M) \xrightarrow{\epsilon} k$$

is a nonsingular bilinear form. If $k = \mathbb{Z}/2$ the orientation assumption can be dropped.

In this proposition the map $\epsilon : \mathbb{H}_q(M) \to k$ is zero unless $q = -d$, in which case it is the isomorphism

$$\mathbb{H}_{-d}(M) = H_0(M) \cong k.$$

Such an algebraic structure, namely a commutative algebra A together with a map $\epsilon : A \to k$ making the pairing $\langle a, b\rangle = \epsilon(a \cdot b)$ a nondegenerate bilinear form, is called a **Frobenius algebra.**

We leave to the reader the exercise of proving the following, (see [Abr96]).

Proposition 1.1.2. *A k-vector space A is a Frobenius algebra if and only if it is a commutative algebra with unit and a cocommutative co-algebra*

$$\Delta : A \to A \otimes A$$

with co-unit $\epsilon : A \to k$, so that Δ is a map of A-bimodules.

Intersection theory can also be realized by the "Thom collapse" map. Consider again the embedding of compact manifolds, $e : P \hookrightarrow M$, and extend e to a tubular neighborhood, $P \subset \eta_e \subset M$. Consider the projection map,

$$\tau_e : M \to M/(M - \eta_e). \tag{1.1}$$

Notice that $M/(M - \eta_e)$ is the one point compactification of the tubular neighborhood, $M/(M - \eta_e) \cong \eta_e \cup \infty$. Furthermore, by the tubular neighborhood theorem, this space is homeomorphic to the Thom space P^{ν_e} of the normal bundle, $\nu_e \to P$,

$$M/(M - \eta_e) \cong \eta_e \cup \infty \cong P^{\nu_e}.$$

So the Thom collapse map can be viewed as a map,

$$\tau_e : M \to P^{\nu_e}. \tag{1.2}$$

Then the homology intersection map $e_!$ is equal to the composition,

$$e_! : H_q(M) \xrightarrow{(\tau_e)_*} H_q(P^{\nu_e}) \cong H_{q-k}(P) \tag{1.3}$$

where the last isomorphism is given by the Thom isomorphism theorem. In fact this description of the umkehr map $e_!$ shows that it can be defined in *any* generalized homology theory, for which there exists a Thom isomorphism for the normal bundle. This is an orientation condition. In these notes we will usually restrict our attention to ordinary homology, but intersection theories in such (co)homology theories as K-theory and cobordism theory are very important as well.

1.2 The Chas-Sullivan loop product

The Chas-Sullivan "loop product" in the homology of the free loop space of a closed oriented d-dimensional manifold,

$$\mu : H_p(LM) \otimes H_q(LM) \to H_{p+q-d}(LM) \tag{1.4}$$

is defined as follows.

Let $\mathrm{Map}(8, M)$ be the mapping space from the figure 8 (i.e the wedge of two circles) to the manifold M. As mentioned above, the maps are required to be piecewise smooth (see [CJ02]). Notice that $\mathrm{Map}(8, M)$ can be viewed as the subspace of $LM \times LM$ consisting of those pairs of loops that agree at the basepoint $1 \in S^1$. In other words, there is a pullback square

$$\begin{array}{ccc} \mathrm{Map}(8, M) & \xrightarrow{e} & LM \times LM \\ {\scriptstyle ev}\big\downarrow & & \big\downarrow{\scriptstyle ev \times ev} \\ M & \xrightarrow[\Delta]{} & M \times M \end{array} \tag{1.5}$$

where $ev : LM \to M$ is the fibration given by evaluating a loop at $1 \in S^1$. In fact, it can be shown that ev is a locally trivial fiber bundle [Kli82]. The map $ev : \mathrm{Map}(8, M) \to M$ evaluates the map at the crossing point of the figure 8. Since $ev \times ev$ is a fibre bundle, $e : \mathrm{Map}(8, M) \hookrightarrow LM \times LM$ can be viewed as a codimension d embedding, with normal bundle $ev^*(\nu_\Delta) \cong ev^*(TM)$.

The basic Chas-Sullivan idea, is to take a chain $c \in C_p(LM \times LM)$ that is transverse to the submanifold $\mathrm{Map}(8, M)$ in an appropriate sense, and take the intersection to define a chain $e_!(c) \in C_{p-d}(\mathrm{Map}(8, M))$. This will allow the definition of a map in homology, $e_! : H_*(LM \times LM) \to H_{*-d}(\mathrm{Map}(8, M))$. The striking thing about the Chas-Sullivan construction is that this umkehr map exists in the absence of Poincaré duality in this infinite dimensional context.

As was done in [CJ02], one can also use the Thom collapse approach to define the umkehr map in this setting. They observed that the existence of this pullback diagram of fiber bundles, means that there is a natural tubular neighborhood of the embedding, $e : \mathrm{Map}(8, M) \to LM \times LM$, namely the inverse image of a tubular neighborhood of the diagonal embedding, $\Delta : M \to M \times M$. That is, $\eta_e = (ev \times ev)^{-1}(\eta_\Delta)$. Because ev is a locally trivial fibration, the tubular neighborhood η_e is homeomorphic to the total space of the normal bundle $ev^*(TM)$. This induces a homeomorphism of the quotient space to the Thom space,

$$(LM \times LM)/((LM \times LM) - \eta_e) \cong (\mathrm{Map}(8, M))^{ev^*(TM)}. \tag{1.6}$$

Combining this homeomorphism with the projection onto this quotient space, defines a Thom-collapse map

$$\tau_e : LM \times LM \to \mathrm{Map}(8, M)^{ev^*(TM)}. \tag{1.7}$$

For ease of notation, we refer to the Thom space of the pullback bundle, $ev^*(TM) \to \mathrm{Map}(8, M)$ as $(\mathrm{Map}(8, M))^{TM}$.

Notice that if h_* is any generalized homology theory that supports an orientation of M (i.e the tangent bundle TM), then one can define an umkehr map,

$$e_! : h_*(LM \times LM) \xrightarrow{\tau_e} h_*((\mathrm{Map}(8, M))^{TM}) \xrightarrow{\cap u} h_{*-d}(\mathrm{Map}(8, M)) \tag{1.8}$$

where $u \in h^d((\mathrm{Map}(8, M))^{TM})$ is the Thom class given by the orientation.

Chas and Sullivan also observed that given a map from the figure 8 to M then one obtains a loop in M by starting at the intersection point, traversing the top loop of the 8, and then traversing the bottom loop. This defines a map

$$\gamma : \mathrm{Map}(8, M) \to LM.$$

Thought of in a slightly different way, the pullback diagram 1.5 says that we can view $\mathrm{Map}(8, M)$ as the fiber product $\mathrm{Map}(8, M) \cong LM \times_M LM$, as was done in [CJ02]. The map γ defines a multiplication map, which by abuse of notation we also call $\gamma : LM \times_M LM \to LM$. This map extends the usual multiplication in the based loop space, $\gamma : \Omega M \times \Omega M \to \Omega M$. In fact if one may view $ev : LM \to M$ as a fiberwise H-space (actually an H-group), which is to say an H-group in the category of spaces over M. It is actually an A_∞ space in this category, coming from the A_∞ structure of the multiplication in ΩM, which is the fiber of $ev : LM \to M$. This aspect of the theory is studied further in [Gru05].

Chas and Sullivan also observed that the multiplication $\gamma : \mathrm{Map}(8, M) \to LM$ is homotopy commutative, and indeed there is a canonical, explicit homotopy. In the formulas that follow, we identify $S^1 = \mathbb{R}/\mathbb{Z}$. Now as above, consider $\mathrm{Map}(8, M)$ as a subspace of $LM \times LM$, and suppose $(\alpha, \beta) \in \mathrm{Map}(8, M)$. We consider, for each $t \in [0, 1]$ a loop $\gamma_t(\alpha, \beta)$, which starts at $\beta(-t)$, traverses the arc between $\beta(-t)$ and $\beta(0) = \alpha(0)$, then traverses the loop defined by α, and then finally traverses the arc between $\beta(0)$ and $\beta(-t)$. A formula for $\gamma_t(\alpha, \beta)$ is given by

$$\gamma_t(\alpha, \beta)(s) = \begin{cases} \beta(2s - t), & \text{for } 0 \leq s \leq \frac{t}{2} \\ \alpha(2s - t), & \text{for } \frac{t}{2} \leq s \leq \frac{t+1}{2} \\ \beta(2s - t), & \text{for } \frac{t+1}{2} \leq s \leq 1. \end{cases} \tag{1.9}$$

One sees that $\gamma_0(\alpha, \beta) = \gamma(\alpha, \beta)$, and $\gamma_1(\alpha, \beta) = \gamma(\beta, \alpha)$.

The Chas-Sullivan product in homology is defined by composing the umkehr map $e_!$ with the multiplication map γ.

Definition 1.2.1. Define the loop product in the homology of a loop space to be the composition

$$\mu : H_*(LM) \otimes H_*(LM) \to H_*(LM \times LM) \xrightarrow{e_!} H_{*-d}(\mathrm{Map}(8, M)) \xrightarrow{\gamma_*} H_{*-d}(LM).$$

Recall that the umkehr map $e_!$ is defined for any generalized homology theory h_* supporting an orientation. Now suppose that in addition, h_* is a multiplicative theory. That is, the corresponding cohomology theory h^* has a cup product, or more precisely, h_* is represented by a *ring* spectrum. Then there is a loop product in $h_*(LM)$ as well,

$$\mu : h_*(LM) \otimes h_*(LM) \to h_{*-d}(LM).$$

In order to accommodate the change in grading, one defines

$$\mathbb{H}_*(LM) = H_{*+d}(LM).$$

Using the naturality of the umkehr map (i.e the naturality of the Thom collapse map) as well as the homotopy commutativity of the multiplication map γ, the following is proved in [CS99].

Theorem 1.2.1. *Let M be a compact, closed, oriented manifold. Then the loop product defines a map*

$$\mu_* : \mathbb{H}_*(LM) \otimes \mathbb{H}_*(LM) \to \mathbb{H}_*(LM)$$

making $\mathbb{H}_*(LM)$ an associative, commutative algebra. Furthermore, the evaluation map $ev : LM \to M$ defines an algebra homomorphism from the loop algebra to the intersection ring,

$$ev_* : \mathbb{H}_*(LM) \to \mathbb{H}_*(M).$$

As was shown in [CJ02], this structure also applies to $h_*(LM)$, where h_* is any multiplicative generalized homology theory which supports an orientation of M.

1.3 The Batalin-Vilkovisky structure and the string bracket

One aspect of the loop space LM that hasn't yet been exploited is the fact there is an obvious circle action

$$\rho : S^1 \times LM \longrightarrow LM \tag{1.10}$$

defined by $\rho(t, \alpha)(s) = \alpha(t + s)$. The purpose of this section is to describe those constructions of Chas and Sullivan [CS99] that exploit this action.

The existence of the S^1-action defines an operator

$$\begin{aligned}\Delta : H_q(LM) &\to H_{q+1}(LM)\\ \theta &\to \rho_*(e_1 \otimes \theta)\end{aligned}$$

where $e_1 \in H_1(S^1) = \mathbb{Z}$ is the generator. Notice that if we apply this operator twice, then $\Delta^2(\theta) = \rho_*(e_1^2 \otimes \theta)$, where the product structure in $H_*(S^1)$ is the Pontrjagin algebra structure, induced by the group structure of S^1. But obviously $e_1^2 = 0$ in this algebra, and so the operator Δ has the property that

$$\Delta^2 = 0.$$

By regrading, this operator may be viewed as a degree one operator on the loop homology algebra,

$$\Delta : \mathbb{H}_*(LM) \to \mathbb{H}_{*+1}(LM)$$

and the following was proved in [CS99].

Theorem 1.3.1. *The pair $(\mathbb{H}_*(LM), \Delta)$ is a Batalin-Vilkovisky (BV) algebra. That is,*

1. *$\mathbb{H}_*(LM)$ is a graded, commutative algebra,*
2. *$\Delta \circ \Delta = 0$, and*
3. *the binary operator defined by the deviation from Δ being a derivation,*

$$\{\phi, \theta\} = (-1)^{|\phi|}\Delta(\phi \cdot \theta) - (-1)^{|\phi|}\Delta(\phi) \cdot \theta - \phi \cdot \Delta(\theta)$$

is a derivation in each variable.

Now a formal argument given in [CS99] shows that the degree one binary operator $\{,\} : \mathbb{H}_*(LM) \times \mathbb{H}_*(LM) \to \mathbb{H}_*(LM)$ described above satisfies the (graded) Jacobi identities, and so makes $\mathbb{H}_*(LM)$ into a graded Lie algebra. Such a combination of being a graded, commutative algebra, as well as a Lie algebra which is a derivation in each variable, is called a Gerstenhaber algebra. So the loop homology algebra has this structure as well.

Chas and Sullivan also gave another description of the bracket $\{\phi, \theta\}$. We give a variation of their description, which is more homotopy theoretic.

Let $\mathcal{P} \subset S^1 \times LM \times LM$ be the space

$$\mathcal{P} = \{(t, \alpha, \beta) : \alpha(0) = \beta(t)\}.$$

Notice that there is a diffeomorphism

$$h : S^1 \times \mathrm{Map}(8, M) \to \mathcal{P}$$

defined by

$$h(t, (\alpha, \beta)) = (t, \alpha, \beta_t) \tag{1.11}$$

where $\beta_t(s) = \beta(s - t)$.

Notice furthermore there is a pullback square of fibrations,

$$\begin{array}{ccc} \mathcal{P} & \xrightarrow[\subset]{u} & S^1 \times LM \times LM \\ {\scriptstyle \epsilon}\downarrow & & \downarrow{\scriptstyle \epsilon} \\ M & \xrightarrow[\Delta]{} & M \times M \end{array} \tag{1.12}$$

where $\epsilon : S^1 \times LM \times LM \to M \times M$ is given by $(t, \alpha, \beta) \to (\alpha(0), \beta(t))$.

Notice that this pullback diagram has a $\mathbb{Z}/2$-action induced by the diagonal action on $S^1 \times LM \times LM$ given by the antipodal map on S^1, and the permutation action on $LM \times LM$. The action on $M \times M$ is also the permutation action under which $\Delta(M)$ is the fixed point set. The diffeomorphism $h : S^1 \times \mathrm{Map}(8, M) \to \mathcal{P}$ is equivariant, where $\mathbb{Z}/2$ acts antipodally on S^1 and permutes the two components of the figure 8.

Like in the previous section, the Thom collapse map may be defined in this situation and we get a map

$$\tau_u : S^1 \times LM \times LM \longrightarrow \mathcal{P}^{TM}.$$

Using the Thom isomorphism this defines an umkehr map,

$$u_! : H_*(S^1 \times LM \times LM) \to H_{*-d}(\mathcal{P}) \cong H_{*-d}(S^1 \times \mathrm{Map}(8, M)).$$

Because of the equivariance, this descends to gives a map on the homology of the orbits,

$$u_! : H_*(S^1 \times_{\mathbb{Z}/2} LM \times LM) \to H_{*-d}(S^1 \times_{\mathbb{Z}/2} \mathrm{Map}(8, M)).$$

This in turn defines a homomorphism

$$\nu_* : H_*(LM \times LM) \to H_{*+1-d}(S^1 \times_{\mathbb{Z}/2} \mathrm{Map}(8, M))$$

by

$$\nu_*(\phi \otimes \theta) = u_!(e_1 \otimes (\phi \otimes \theta - (-1)^{(|\phi|+1)(|\theta|+1)} \theta \otimes \phi). \tag{1.13}$$

We observe that since the maps $u_!$ and ν_* are defined in terms of the Thom collapse map τ_u, they can be defined in any generalized homology theory that supports an orientation of M.

Now let $G : [0, 1] \times \mathrm{Map}(8, M) \to LM$ be the homotopy given by (1.9). By identifying $[0, 1]$ with the upper semicircle of S^1, G defines a map

$$G : S^1 \times_{\mathbb{Z}/2} \mathrm{Map}(8, M) \to LM.$$

By composing we then have an operation,

$$G_* \circ \nu_* : H_*(LM \times LM) \to H_{*+1-d}((S^1 \times_{\mathbb{Z}/2} \mathrm{Map}(8, M)) \to H_{*+1-d}(LM).$$

This operation is easily seen to be a global description of the bracket operation defined in definition 4.1 of [CS99]. Corollary 5.3 of [CS99] then gives the following.

Theorem 1.3.2. *$(G_* \circ \nu_*)(\phi \otimes \theta) = \{\phi, \theta\}$.*

Notice that from this global description, which defines the Batalin-Vilkovisky structure ultimately in terms of the Thom collapse map, we have the following generalization of Theorem 1.3.1, originally proved in a somewhat different way by Cohen and Jones in [CJ02].

Theorem 1.3.3. *Let h_* be any multiplicative generalized homology theory that supports an orientation of M. Then $h_*(LM)$ is a Batalin-Vilkovisky algebra.*

We now turn to the effect of the Batalin-Vilkovisky structure on the equivariant homology, $H_*^{S^1}(LM)$. Chas and Sullivan refer to this as the "string homology" of M. It is the homology of the homotopy orbit space,

$$H_*^{S^1}(LM) = H_*(ES^1 \times_{S^1} LM).$$

We view this homotopy orbit space as the "space of closed strings" in M for the following reason.

Consider the space of embeddings of S^1 into an infinite dimensional Euclidean space, $Emb(S^1, \mathbb{R}^\infty)$. This is a contractible space with an obvious free action of S^1, given by reparameterization of the embeddings. Indeed this action extends to a free action of the homotopy equivalent, but much larger group of orientation preserving diffeomorphisms, $Diff^+(S^1)$. So the homotopy orbit space $ES^1 \times_{S^1} LM$ is homotopy equivalent to the orbit space, $Emb(S^1, \mathbb{R}^\infty) \times_{Diff^+(S^1)} LM$ which can be described on the point set level as follows:

$$Emb(S^1, \mathbb{R}^\infty) \times_{Diff^+(S^1)} LM = \{(S, f)\}$$

where $S \subset \mathbb{R}^\infty$ is a closed, oriented, connected one dimensional submanifold of $\mathbb{R}^\infty$, and $f : S \to M$ is a continuous map. In other words, this orbit space is the space of oriented closed curves (closed strings) in M. Thus the equivariant homology, $H_*^{S^1}(LM)$ is the homology of the space of closed strings in M.

Now consider the principal S^1 bundle,

$$S^1 \to ES^1 \times LM \to ES^1 \times_{S^1} LM. \tag{1.14}$$

This gives rise to the Gysin sequence in homology,

$$\to \cdots H_q(LM) \xrightarrow{\iota_*} H_q^{S^1}(LM) \xrightarrow{j_*} H_{q-2}^{S^1}(LM) \xrightarrow{\tau_*} H_{q-1}^{S^1}(LM) \xrightarrow{\iota_*} \cdots \tag{1.15}$$

We recall that the connecting homomorphism $\tau : H_k^{S^1}(LM) \to H_{k+1}(LM)$ is induced by the S^1-transfer map for this principal bundle , which is defined via a Thom collapse map. This is a stable map (map of suspension spectra),

$$\tau : \Sigma^\infty(\Sigma(ES^1 \times_{S^1} X)_+) \to \Sigma^\infty(X_+)$$

which exists for any space X with an S^1-action [BG75]. Here Σ^∞ refers to the suspension spectrum, and Y_+ is Y with a disjoint basepoint. The map τ_* in the Gysin sequence is homomorphism induced by τ in homology.

We note that Chas and Sullivan denoted the homomorphism ι_* by E, standing for "erase", and the homomorphism τ_* by M, standing for "mark". The motivation for this terminology is that one might consider the loop space LM as the space of "marked" closed strings, since the space of "markings" (i.e choices of marked point) on a closed, oriented, connected one dimensional manifold S, is homeomorphic to S^1, which is homotopy equivalent to the space of parameterizations of S by a diffeomorphism from the circle, $S^1 \xrightarrow{\cong} S$. Thus the homomorphism $\iota_* = E$ can be viewed as erasing the marking, and the transfer homomorphism, $\tau_* = M$ can be viewed as taking the map in homology induced by taking all possible markings on $S \subset \mathbb{R}^\infty$. Since the space of markings is one dimensional, this accounts for the dimension shift in the homomorphism τ_*.

Using these maps, Chas and Sullivan define an operator

$$\begin{aligned} [\,,\,] : H_q^{S^1}(LM) \times H_r^{S^1}(LM) &\xrightarrow{\tau_* \times \tau_*} H_{q+1}(LM) \times H_{r+1}(LM) \\ &\xrightarrow{\circ} H_{q+r+2-d}(LM) \xrightarrow{\iota_*} H_{q+r+2-d}^{S^1}(LM). \end{aligned} \tag{1.16}$$

where "$\circ$" is the loop product described above. Using just the formal properties of the Batalin-Vilkovisky structure on $H_*(LM)$, Chas and Sullivan proved the following in [CS99].

Theorem 1.3.4. *The operator*

$$[\,,\,] : H_q^{S^1}(LM) \times H_r^{S^1}(LM) \longrightarrow H_{q+r+2-d}^{S^1}(LM)$$

gives the string homology $H_*^{S^1}(LM)$ the structure of a graded Lie algebra of degree $(2-d)$.

We note that since the transfer map τ comes from a map of spectra, there is a corresponding string bracket on $h_*^{S^1}(LM) = h_*(ES^1 \times_{S^1} LM)$ for h_* any multiplicative generalized homology theory that supports an orientation of M.

The string bracket gives the equivariant homology a very rich structure. For example, if M is a surface, the bracket restricts to define a Lie algebra structure on the vector space generated by the path components of closed curves in a surface,

$$[\,,\,] : H_0^{S^1}(LM^2) \times H_0^{S^1}(LM^2) \longrightarrow H_0^{S^1}(LM^2).$$

This bracket operation was originally discovered by Wolpert [Wol83] and Goldman [Gol86] and is highly nontrivial. It has been studied in more depth by Chas in [Cha01].

1.4 A stable homotopy point of view

In this section we describe a homotopy theoretic realization of the loop product due to Cohen and Jones [CJ02]. This takes the form that the Thom spectrum of a certain virtual bundle over the loop space is a ring spectrum, which in homology realizes the loop product.

Recall that in the definition of the loop product (1.2.1), an essential ingredient was the Thom collapse map (1.7)

$$\tau_e : LM \times LM \to \mathrm{Map}(8, M)^{ev^*(TM)}.$$

The general Thom collapse map (1.2) can be modified by twisting with bundles in the following manner.

Given $e : N \hookrightarrow M$, and $\zeta \to M$ vector bundle, consider the embedding of total spaces,

$$\begin{array}{ccc} e^*\zeta & \xrightarrow[\hookrightarrow]{e_\zeta} & \zeta \\ \downarrow & & \downarrow \\ N & \xrightarrow[\hookrightarrow]{e} & M \end{array}$$

The tubular neighborhood of e_ζ is homeomorphic to total space of $e^*\zeta \oplus \eta_e$. So we get a Thom collapse map

$$\tau : M^\zeta \to N^{e^*\zeta \oplus \eta_e}$$

This construction can be carried out even if ζ is a *virtual* bundle, $\zeta = \gamma_1 - \gamma_2$, where γ_i's are vector bundles. We view a virtual bundle as an element of K-theory, and there is a dimension homomorphism, $\dim : K(M) \to \mathbb{Z}$, where $\dim(\zeta) = \dim(\gamma_1) - \dim(\gamma_2)$. So the dimension is possibly negative.

In this setting, the Thom isomorphism still holds, $H^q(M) \cong H^{q+\dim\zeta} M^\zeta$, where again, $\dim \zeta$ might be negative.

Example

Let $\zeta \to M$ be a vector bundle. Consider the product bundle $\zeta \times \zeta \to M \times M$, and embedding $\Delta : M \hookrightarrow M \times M$. The Thom collapse is a map of spectra,

$$\tau : M^\zeta \wedge M^\zeta \longrightarrow M^{2\zeta \oplus TM}$$

Now take ζ to be the virtual bundle, $\zeta = -TM$. The Thom collapse is then a map

$$\tau : M^{-TM} \wedge M^{-TM} \longrightarrow M^{-TM} \tag{1.17}$$

This defines a ring structure on M^{-TM} which was studied in detail in [Coh04b]. We describe this structure more explicitly as follows.

Let $e : M \to \mathbb{R}^L$ be an embedding, with η_e its normal bundle. Then there is an equivalence between the L-fold suspension of the Thom spectrum M^{-TM} and the Thom space of the normal bundle, M^{η_e},

$$\Sigma^L M^{-TM} \simeq M^{\eta_e}.$$

For $\epsilon > 0$, let ν_e^ϵ be the ϵ-tubular neighborhood: $= \{y \in \mathbb{R}^L : d(y, e(M)) < \epsilon\}$. Then for ϵ sufficiently small,

$$M^{\eta_e} \cong \mathbb{R}^L/(\mathbb{R}^L - \nu_e^\epsilon).$$

Let $B_\epsilon(0)$ be the ball of radius ϵ around the origin in $\mathbb{R}^L$. In the 1930's, Alexander considered the map

$$\begin{aligned}(\mathbb{R}^L - \nu_e^\epsilon) \times M &\longrightarrow \mathbb{R}^L - B_\epsilon(0) \simeq S^{L-1} \\ (v, x) &\longrightarrow v - e(x)\end{aligned}$$

This defines a map in homology,

$$H_{L-q-1}(\mathbb{R}^L - \nu_e^\epsilon) \otimes H_q(M) \longrightarrow H_{L-1}S^{L-1} \cong \mathbb{Z}$$

or by taking the adjoint,

$$H_q(M) \longrightarrow H^{L-q-1}(\mathbb{R}^L - \nu_e^\epsilon)$$

This is the famous "Alexander duality" isomorphism.

In early 1960's, Atiyah [Ati61] considered the same map defined on the quotient spaces,

$$\begin{aligned}\mathbb{R}^L \times M/((\mathbb{R}^L - \nu_e^\epsilon) \times M) = M^{\eta_e} \wedge M_+ &\to \mathbb{R}^L/(\mathbb{R}^L - B_\epsilon(0)) \\ &\cong S^L \\ (v, x) &\longrightarrow v - e(x)\end{aligned}$$

This yields a map

$$\alpha : M^{\eta_e} \to \mathrm{Map}(M_+, S^L).$$

On the spectrum level, α induces a map

$$\alpha : M^{-TM} \to F(M_+, S),$$

where $F(M_+, S)$ is the space of stable maps from M_+ to the sphere spectrum S. As a spectrum, its kth space is given by $\mathrm{Map}(M_+, S^k)$.

The following theorem, proved by Atiyah in [Ati61] says that this map is a homotopy equivalence.

Theorem 1.4.1. *For a closed manifold M, the map $\alpha : M^{-TM} \to F(M_+, S)$, is homotopy equivalence from the Thom spectrum of $-TM$ to the Spanier-Whitehead dual, of M_+.*

The Spanier-Whitehead dual, $F(M_+, S)$ is clearly a ring spectrum. Its multiplication is dual to the diagonal map $\Delta : M \to M \times M$. Since this diagonal map is cocommutative, this says that the ring structure on $F(M_+, S)$ also has commutativity properties. In recent years, symmetric monoidal categories of spectra have been developed ([EKMM97], [HSS00]) in which the appropriate coherence issues regarding homotopy commutativity can be addressed. In [Coh04b], Cohen used the notion of symmetric spectrum developed in [HSS00] and proved the following.

Theorem 1.4.2. *α is an equivalence of commutative, symmetric ring spectra and of bimodules over $F(M_+, S)$.*

That is, the ring structure on M^{-TM} induced by the Thom collapse map (1.17), can be rigidified to give a commutative ring structure which coincides, via the classical map of Alexander, to the commutative ring structure on $F(M_+, S)$ given by the dual of the diagonal map.

Now as seen above, the ring map (1.17) is determined by the Thom collapse map for the pullback diagram of virtual bundles,

$$\begin{array}{ccc} -2TM & \xrightarrow{\Delta} & -TM \times -TM \\ \downarrow & & \downarrow \\ M & \underset{\Delta}{\xhookrightarrow{}} & M \times M. \end{array}$$

We now pull this diagram of virtual bundles back, using the evaluation map, $ev : LM \to M$, to the total spaces of pullback square (1.5).

$$\begin{array}{ccc} ev^*(-2TM) & \xrightarrow{e} & ev^*(-TM) \times ev^*(-TM) \\ \downarrow & & \downarrow \\ \mathrm{Map}(8, M) & \underset{e}{\xhookrightarrow{}} & LM \times LM \end{array}$$

This defines a Thom collapse map,

$$\tau_e : LM^{-TM} \wedge LM^{-TM} \to \mathrm{Map}(8, M)^{-TM}$$

which yields a product

$$\mu : LM^{-TM} \wedge LM^{-TM} \xrightarrow{\tau_e} \mathrm{Map}(8, M)^{-TM} \xrightarrow{\gamma} LM^{-TM}.$$

In [CJ02] Cohen and Jones proved the following.

Theorem 1.4.3. *For any closed manifold M, LM^{-TM} is a ring spectrum. If M is oriented, then the Thom isomorphism in homology,*

$$H_*(LM^{-TM}) \cong H_{*+n}LM \cong \mathbb{H}_*(LM)$$

is an isomorphism of commutative graded algebras. The analogous statement holds for any multiplicative generalized homology theory h_* *that supports an orientation of* M*. Furthermore the evaluation map*

$$ev : LM^{-TM} \to M^{-TM}$$

is a map of ring spectra. Also, if one considers pullback diagram

$$\begin{array}{ccc} \Omega M & \xrightarrow{\hookrightarrow} & LM \\ \downarrow & & \downarrow{\scriptstyle ev_0} \\ point & \xrightarrow[\hookrightarrow]{} & M \end{array}$$

then the induced Thom collapse map

$$LM^{-TM} \to \Sigma^\infty(\Omega M_+)$$

is a map of ring spectra.

The existence of ring structures on Thom spectra of more general fiber products was examined by Klein in [Kle03].

This theorem says that even on the stable homotopy level, the string topology structure on the loop space is compatible with both the intersection product on M, and Pontrjagin product on ΩM.

This compatibility was strengthened by Cohen, Jones, and Yan in [CJY03] for calculational purposes. Namely, consider Serre spectral sequence for the fibration

$$\Omega M \to LM \xrightarrow{ev} M$$

Assume M is simply connected. Then the E_2 term of the spectral sequence is given by $E_2^{*,*} = H_*(M, H_*(\Omega M))$, and the spectral sequence converges to $H_*(LM)$. By applying Poincaré duality, one gets a second quadrant spectral sequence

$$E_2^{-s,t} = H^s(M; H_t(\Omega M)) \Rightarrow H_{n-s+t}(LM) = \mathbb{H}_{t-s}(LM).$$

Notice that this E_2-term is an algebra, using cup product in cohomology and the Pontrjagin product on the coefficients. It was shown in [CJY03] that this spectral sequence is multiplicative. That is, each $E_r^{*,*}$ is a graded ring, and the differentials are derivations. It converges multiplicatively to the loop product on $\mathbb{H}_*(LM)$. Cohen, Jones, and Yan then used this spectral sequence to calculate the ring structure in $\mathbb{H}_*(L\mathbb{CP}^n)$ and $\mathbb{H}_*(LS^m)$ for all n and m, and demonstrated what a rich structure this algebra detects.

1.5 Relation to Hochschild cohomology

In this section we describe an algebraic point of view of the string product. Namely we recall how Hochschild cohomology of the cochains of a manifold, $H^*(C^*(M), C^*(M))$ was shown in [CJ02] to be isomorphic as rings to $\mathbb{H}_*(LM)$. This relationship is an outgrowth of the cosimplicial model of the loop space and its relationship to Hochschild homology established by Jones in [Jon87].

We begin by recalling the basic construction. Consider the standard simplicial decomposition of the circle, S^1 which has one 0-simplex and one nondegenerate 1-simplex.

In a simplicial set $\mathcal{S}_*$, there are face maps $\partial_i : \mathcal{S}_k \to \mathcal{S}_{k-1}$, $i = 0, \cdots k$, and degeneracy maps $\sigma_j : \mathcal{S}_k \to \mathcal{S}_{k+1}$, $j = 0, \cdots, k$. So S^1_* is generated with respect to the degeneracies by the one zero simplex and one nondegenerate one-simplex. Using the relations satisfied by the face and degeneracy operators, it turns out that the set of k-simplices, S^1_k is a set with $k+1$ elements.

$$S^1_k = \{\mathbf{k+1}\}$$

All the simplices are degenerate if $k > 1$.

Now recall the coface and codegeneracy maps on the standard simplices:

$$\begin{aligned} d^i &: \Delta^{k-1} \to \Delta^k \quad i = 0, \cdots, k \\ s^j &: \Delta^{k+1} \to \Delta^k \quad j = 0, \cdots, k. \end{aligned}$$

We then have the resulting homeomorphism of the geometric realization

$$S^1 \cong \bigcup_k \Delta^k \times S^1_k / \sim \quad = \bigcup_k \Delta^k \times \{\mathbf{k+1}\} / \sim$$

where $(s^j(t), x) \sim (t, \sigma_j(x))$, $(d^i(t), y) \sim (t, d_i(y))$.

This gives an embedding of the loop space,

$$\begin{aligned} f = \prod_k f_k : LX = \mathrm{Map}(S^1, X) &= \mathrm{Map}(\bigcup_k \Delta^k \times \{\mathbf{k+1}\} / \sim, \mathrm{X}) \\ &\subset \mathrm{Map}(\coprod_k \Delta^k \times \{\mathbf{k+1}\}; \mathrm{X}) \\ &= \prod_k \mathrm{Map}(\Delta^k; X^{k+1}) \end{aligned} \tag{1.18}$$

The image of this embedding are those sequences of maps that commute with the coface and codegeneracy operators.

The component maps $f_k : LX \to \mathrm{Map}(\Delta^k, X)$ can be described explicitly as follows.

$$\begin{aligned} f_k : \Delta^k \times LX &\longrightarrow X^{k+1} \\ (0 \leq t_1 \leq \cdots \leq t_k \leq 1; \gamma) &\to (\gamma(t_1), \cdots \gamma(t_k), \gamma(1)). \end{aligned}$$

If we consider the homomorphism on singular cochains induced by f_k, and then take the cap product with the canonical k-simplex in the chains of Δ^k, we get a diagram, which in [Jon87] was shown to commute:

$$\begin{array}{ccc} C^{*-k}(LX) & \xleftarrow{f_k^*} & C^*(X)^{\otimes k+1} \\ \delta \downarrow & & \downarrow total\, diff \\ C^{*-k+1}(LX) & \xleftarrow[f_{k+1}^*]{} & C^*(X)^{\otimes k} \end{array}$$

Here the right hand vertical map is the total differential in the Hochschild chain complex. Recall that the Hochschild chain complex of a differential graded algebra, with coefficients in a bimodule C is the complex $CH_*(A, C)$:

$$\longrightarrow \cdots \xrightarrow{\partial} C \otimes A^{\otimes n} \xrightarrow{\partial} C \otimes A^{\otimes n-1} \xrightarrow{\partial} \cdots$$

where ∂ is the total differential, given by the sum of the internal differential on $A^{\otimes n} \otimes C$ and the Hochschild boundary operator

$$\begin{aligned} b(c \otimes a_1 \otimes \cdots \otimes a_n) = & \; c \cdot a_1 \otimes a_2 \otimes \cdots \otimes a_n \\ + & \sum_{i=1}^{n-1} (-1)^i c \otimes a_1 \otimes \cdots a_i \cdot a_{i+1} \otimes \cdots \otimes a_n \\ + & \; (-1)^n a_n \cdot c \otimes a_1 \otimes \cdots \otimes a_{n-1}. \end{aligned}$$

In the case above, we are considering $CH_*(C^*(X), C^*(X))$, the Hochschild complex. The following was proved by Jones in [Jon87].

Theorem 1.5.1. *For simply connected X,*

$$f^* : CH_*(C^*(X), C^*(X)) \longrightarrow C^*(LX)$$

is a chain homotopy equivalence. It therefore induces an isomorphism

$$f^* : H_*(C^*(X), C^*(X)) \xrightarrow{\cong} H^*(LX).$$

Now the loop product lives in the homology $H_*(LM)$, so to get a model for this, we dualize the Hochschild complex, and we get a complex

$$\longrightarrow \cdots \longrightarrow \mathrm{Hom}(C^*(X)^{\otimes q}, k) \xrightarrow{\delta} \mathrm{Hom}(C^*(X)^{\otimes k+1}, k) \xrightarrow{\delta} \cdots$$

which computes the homology $H_*(LX; k)$. But with respect to the obvious identification, $\mathrm{Hom}(C^*(X)^{\otimes q+1}, k) \cong \mathrm{Hom}(C^*(X)^{\otimes q}; C_*(X))$, this complex is the Hochschild cochain complex of $C^*(X)$ with coefficients in the bimodule $C_*(X)$. We therefore have the following corollary.

Corollary 1.5.2. *For simply connected X, there is an isomorphism*

$$f_* : H_*(LX) \xrightarrow{\cong} H^*(C^*(X); C_*(X)).$$

Now let $X = M^n$ be a simply connected, oriented, closed manifold. Notice that the following diagram commutes:

$$\begin{array}{ccc} \Delta^q \times LM & \xrightarrow{f_q} & M^{q+1} \\ {\scriptstyle ev}\downarrow & & \downarrow{\scriptstyle p_{q+1}} \\ M & \xrightarrow[=]{} & M \end{array}$$

where p_{q+1} is the projection onto the last factor. This implies we have a map of Thom spectra,

$$f_q : \Delta^q_+ \wedge LM^{-TM} \to M^q_+ \wedge M^{-TM}$$

for each q. It was shown in [CJ02] that as a consequence of the embedding (1.18) and Theorem 1.5.1, we have the following:

Theorem 1.5.3. *For M simply connected, the map f induces an isomorphism of rings,*

$$f_* : \mathbb{H}_*(LM) \cong H_*(LM^{-TM}) \xrightarrow{\cong} H^*(C^*(M); C_*(M^{-TM}))$$

The ring structure of the Hochschild cohomology is given by cup product, where one is using the ring spectrum structure of the Atiyah dual, M^{-TM} to give a ring structure to the coefficients, $C_*(M^{-TM})$.

As a consequence of the fact that the Atiyah duality map α is an equivalence of ring spectra, (Theorem 1.4.2), it was shown in [Coh04b] that α induces an isomorphism,

$$\alpha_* : H^*(C^*(M); C_*(M^{-TM})) \xrightarrow{\cong} H^*(C^*(M), C^*(M)).$$

This then implies the following Theorem [CJ02]:

Theorem 1.5.4. *The composition*

$$\psi : \mathbb{H}_*(LM) \xrightarrow{f_*} H^*(C^*(M); C_*(M^{-TM})) \xrightarrow{\alpha_*} H^*(C^*(M), C^*(M))$$

is an isomorphism of graded algebras.

We end this section with a few comments and observations about recent work in this direction.

Comments

1. Clearly the Hochschild cohomology, $H^*(C^*(M), C*(M))$ only depends on the homotopy type of M. However the definition of the loop product and Batalin-Vilkovisky structure on $\mathbb{H}_*(LM)$ involves intersection theory, and as we've described it above, the Thom collapse maps. These constructions involve the smooth structure of M. Indeed, even the definition of the isomorphism, $\psi : \mathbb{H}_*(LM) \to H^*(C^*(M), C*(M))$ given in [CJ02] involves the Thom collapse map, and therefore the smooth structure of M. Nonetheless, using the Poincaré embedding theory of Klein [Kle99], Cohen, Klein, and Sullivan have recently proved that if h_* is a multiplicative generalized homology theory supporting an orientation of M, and $f : M_1 \to M_2$ is an h_*-orientation preserving homotopy equivalence of simply connected, closed manifolds, then the induced homotopy equivalence of loop spaces, $Lf : LM_1 \to LM_2$ induces an isomorphism of BV-algebras, $(Lf)_* : h_*(LM_1) \to h_*(LM_2)$ [CKS05].

2. Félix, Menichi, and Thomas [FMT02] proved that the Hochschild cohomology $H^*(C^*(M), C^*(M))$ is a BV algebra. It is expected that the ring isomorphism

$$\psi : \mathbb{H}_*(LM) \to H^*(C^*(M), C^*(M))$$

preserves the BV structure. However this has not yet been proved. They also show that a kind of Koszul duality implies that there is an isomorphism of Hochschild cohomologies,

$$H^*(C^*(M), C^*(M)) \cong H^*(C_*(\Omega M), C_*(\Omega M)) \tag{1.19}$$

where the chains on the based loop space $C_*(\Omega M)$ has the Pontrjagin algebra structure. This is significant because of the alternative description of the homology $H_*(LX)$ in terms of Hochschild homology [Goo85]

$$H_*(LX) \cong H_*(C_*(\Omega X), C_*(\Omega X))$$

for any X. (X need not be simply connected in this case.)

3. The topological Hochschild cohomology of a ring spectrum R, $THH^*(R)$ (see [BHM93]) is a cosimplicial spectrum which has the structure of an algebra over the little disk operad by work of McClure and Smith. The cosimplicial model for the suspension spectrum LM^{-TM} constructed by Cohen and Jones [CJ02] is equivalent to $THH^*(M^{-TM})$, which by Atiyah duality (Theorem (1.4.2)) is equivalent to $THH^*(F(M_+, S))$. It was observed originally by Dwyer and Miller, as well as Klein [Kle03] that the topological Hochschild cohomology $THH^*(\Sigma^\infty(\Omega(M_+))$ is also homotopy equivalent to LM^{-TM}.

The spectrum homology of these spectra are given by

$$H_*(THH^*(F(M_+, S))) \cong H^*(C^*(M), C^*(M))$$
$$H_*(THH^*(\Sigma^\infty(\Omega(M_+)))) \cong H^*(C_*(\Omega X), C_*(\Omega X))$$

and so the equivalence,

$$H_*(THH^*(F(M_+, S))) \simeq \Sigma^\infty(LM_+) \simeq H_*(THH^*(\Sigma^\infty(\Omega M_+))$$

realizes, on the spectrum level, the Koszul duality isomorphism (1.19) above.

Chapter 2

The cacti operad

2.1 PROPs and operads

Operads in general are spaces of operations with certain rules on how to compose the operations. In this sense operads are directly related to Lawvere's algebraic theories and represent true objects of universal algebra. However, operads as such appeared in topology in the works of J. P. May [May96], J. M. Boardman and R. M. Vogt [BV73] as a recognition tool for based multiple loop spaces. Stasheff [Sta63] earlier described the first example of an operad, the associahedra, which recognized based loop spaces. About the same time, Gerstenhaber [Ger68], studying the algebra of the Hochschild complex, introduced the notion of a composition algebra, which is equivalent to the notion of an operad of graded vector spaces.

2.1.1 PROP's

We will start with defining the notion of a PROP (=PROducts and Permutations) and think of an operad as certain part of a PROP. However, later we will give an independent definition of an operad.

Definition 2.1.1. A *PROP* is a symmetric monoidal (sometimes called tensor) category whose set of objects is identified with the set $\mathbb{Z}_+$ of nonnegative integers. The monoidal law on $\mathbb{Z}_+$ is given by addition and the associativity transformation α is equal to identity. See the founding fathers' sources, such as, J. F. Adams' book [Ada78] or S. Mac Lane's paper [ML65] for more detail.

Usually, PROP's are enriched over another symmetric monoidal category, that is, the morphisms in the PROP are taken as objects of the other symmetric monoidal category. This gives the notions of a PROP of sets, vector spaces, complexes, topological spaces, manifolds, etc. Examples of PROP's include the following. We will only specify the morphisms, because the objects are already given by the definition.

Example 2.1.1. The *endomorphism PROP* of a vector space V has the space of morphisms $\mathrm{Mor}(m, n) = \mathrm{Hom}(V^{\otimes m}, V^{\otimes n})$. This is a PROP of vector spaces. The composition and tensor product of morphisms are defined as the corresponding operations on linear maps.

Example 2.1.2 (Segal [Seg88]). The *Segal PROP* is a PROP of infinite dimensional complex manifolds. A morphism is defined as a point in the moduli space $\mathcal{P}_{m,n}$ of isomorphism classes of complex Riemann surfaces bounding $m + n$ labeled nonoverlapping holomorphic holes. The surfaces should be understood as compact smooth complex curves, not necessarily connected, along with $m+n$ biholomorphic maps of the closed unit disk to the surface, thought of as holes. The biholomorphic maps are part of the data, which in particular means that choosing a different biholomorphic map for the same hole is likely to change the point in the moduli space. The more precise nonoverlapping condition is that the closed disks in the inputs do not intersect pairwise and the closed disks in the outputs do not intersect pairwise, however, an input and an output disk may have common boundary, but are still not allowed to intersect at an interior point. This technicality brings in the identity morphisms to the PROP, but does not create singular Riemann surfaces by composition. The composition of morphisms in this PROP is given by sewing the Riemann surfaces along the boundaries, using the equation $zw = 1$ in the holomorphic parameters coming from the standard one on the unit disk. The tensor product of morphisms is the disjoint union. This PROP plays a crucial role in Conformal Field Theory, as we will see now.

2.1.2 Algebras over a PROP

We need to define another important notion before we proceed.

Definition 2.1.2. We say that a vector space V is an *algebra over a PROP P*, if a morphism of PROP's from P to the endomorphism PROP of V is given. A morphism of PROP's is a functor respecting the symmetric monoidal structures and also equal to the identity map on the objects.

An algebra over a PROP could have been called a *representation*, but since algebras over operads, which are similar objects, are nothing but familiar types of algebras, it is more common to use the term "algebra."

Example 2.1.3. An example of an algebra over a PROP is a *Conformal Field Theory* (*CFT*), which may be defined (in the case of a vanishing central charge) as an algebra over the Segal PROP. The fact that the functor respects compositions of morphisms translates into the sewing axiom of CFT in the sense of G. Segal. Usually, one also asks for the functor to depend smoothly on the point in the moduli space $\mathcal{P}_{m,n}$. One needs to extend the Segal PROP by a line bundle to cover the case of an arbitrary charge, see [Hua97].

Example 2.1.4 (Sullivan). Another example of an algebra over a PROP is a Lie bialgebra. There is a nice graph description of the corresponding PROP, about which we learned from Sullivan, see [MV03].

2.1.3 Operads

Now we are ready to deal with operads, which formalize the notion of a space of operations, as we mentioned in the introduction to Section 2.1. Informally, an operad is the part $\mathrm{Mor}(n,1)$, $n \geq 0$, of a PROP. Of course, given only the collection of morphisms $\mathrm{Mor}(n,1)$, it is not clear how to compose them. The idea is to take the union of m elements from $\mathrm{Mor}(n,1)$ and compose them with an element of $\mathrm{Mor}(m,1)$. This leads to cumbersome notation and ugly axioms, compared to those of a PROP. However operads are in a sense more basic than the corresponding PROP's; the difference is similar to the difference between Lie algebras and the universal enveloping algebras.

Definition 2.1.3 (May [May96]). An *operad* $\mathcal{O}$ is a collection of sets (vector spaces, complexes, topological spaces, manifolds, ..., objects of a symmetric monoidal category) $\mathcal{O}(n)$, $n \geq 0$, with

1. A composition law:
$$\gamma : \mathcal{O}(m) \otimes \mathcal{O}(n_1) \otimes \cdots \otimes \mathcal{O}(n_m) \to \mathcal{O}(n_1 + \cdots + n_m).$$
2. A right action of the symmetric group Σ_n on $\mathcal{O}(n)$.
3. A unit $e \in \mathcal{O}(1)$.

such that the following properties are satisfied:

1. The composition is associative, *i.e.*, the following diagram is commutative:
$$\begin{array}{ccc} \left\{\begin{array}{l} \mathcal{O}(l) \otimes \mathcal{O}(m_1) \otimes \cdots \otimes \mathcal{O}(m_l) \\ \qquad \otimes\, \mathcal{O}(n_{1,1}) \otimes \cdots \otimes \mathcal{O}(n_{l,n_l}) \end{array}\right\} & \xrightarrow{\mathrm{id} \otimes \gamma^l} & \mathcal{O}(l) \otimes \mathcal{O}(n_1) \otimes \cdots \otimes \mathcal{O}(n_l) \\ {\scriptstyle \gamma \otimes \mathrm{id}} \downarrow & & \downarrow {\scriptstyle \gamma} \\ \mathcal{O}(m) \otimes \mathcal{O}(n_{1,1}) \otimes \cdots \otimes \mathcal{O}(n_{m,n_m}) & \xrightarrow{\gamma} & \mathcal{O}(n) \end{array},$$
where $m = \sum_i m_i$, $n_i = \sum_j n_{i,j}$, and $n = \sum_i n_i$.
2. The composition is equivariant with respect to the symmetric group actions: the groups Σ_m, Σ_{n_1}, ..., Σ_{n_m} act on the left-hand side and map naturally to $\Sigma_{n_1+\cdots+n_m}$, acting on the right-hand side.
3. The unit e satisfies natural properties with respect to the composition: $\gamma(e; f) = f$ and $\gamma(f; e, \ldots, e) = f$ for each $f \in \mathcal{O}(k)$.

   The notion of a *morphism of operads* is introduced naturally.

Remark 1. One can consider *non-Σ operads*, not assuming the action of the symmetric groups. Not requiring the existence of a unit e, we arrive at *nonunital operads*. Do not mix this up with operads with no $\mathcal{O}(0)$, algebras over which (see next section) have no unit. There are also good examples of operads having only $n \geq 2$ components $\mathcal{O}(n)$.

An equivalent definition of an operad may be given in terms of operations $f \circ_i g = \gamma(f; \mathrm{id}, \dots, \mathrm{id}, g, \mathrm{id}, \dots, \mathrm{id})$, $i = 1, \dots, m$, for $f \in \mathcal{O}(m), g \in \mathcal{O}(n)$. Then the associativity condition translates as $f \circ_i (g \circ_j h) = (f \circ_i g) \circ_{i+j-1} h$ plus a natural symmetry condition for $(f \circ_i g) \circ_j h$, when g and h "fall into separate slots" in f, see *e.g.*, [KSV96].

Example 2.1.5 (The Riemann surface and the endomorphism operads). $\mathcal{P}(n)$ is the space of Riemann spheres with $n + 1$ boundary components, *i.e.*, n inputs and 1 output. Another example is the *endomorphism operad of a vector space* V: $\mathcal{E}nd_V(n) = \mathrm{Hom}(V^{\otimes n}, V)$, the space of n-linear mappings from V to V.

2.1.4 Algebras over an operad

Definition 2.1.4. An *algebra over an operad* $\mathcal{O}$ (in other terminology, a *representation of an operad*) is a morphism of operads $\mathcal{O} \to \mathcal{E}nd_V$, that is, a collection of maps

$$\mathcal{O}(n) \to \mathcal{E}nd_V(n) \qquad \text{for } n \geq 0$$

compatible with the symmetric group action, the unit elements, and the compositions. If the operad $\mathcal{O}$ is an operad of vector spaces, then we would usually require the morphism $\mathcal{O} \to \mathcal{E}nd_V$ to be a morphism of operads of vector spaces. Otherwise, we would think of this morphism as a morphism of operads of sets. Sometimes, we may also need a morphism to be continuous or respect differentials, or have other compatibility conditions. We will also consider a nonlinear version of the notion of an operad algebra, which may be defined in any symmetric monoidal category. For example, an $\mathcal{O}$-algebra X in the category of topological spaces would be an operad morphism $\mathcal{O}(n) \to \mathrm{Map}(X^n, X)$ for $n \geq 0$, where Map is the space of continuous maps.

The commutative operad

The *commutative operad* is the operad of k-vector spaces with the nth component $\mathcal{C}omm(n) = k$ for all $n \geq 0$. We assume that the symmetric group acts trivially on k and the compositions are just the multiplication of elements in the ground field k. The term "commutative operad" may seem confusing to some people, but it has been in use for a while. An algebra over the commutative operad is nothing but a commutative associative algebra with a unit, as we see from Exercise 2 below.

Another version of the commutative operad is $\mathcal{C}omm(n) = \{\text{point}\}$ for all $n \geq 0$. This is an operad of sets. It is equivalent to the previous version in the sense that an algebra over it is the same as a commutative associative unital algebra.

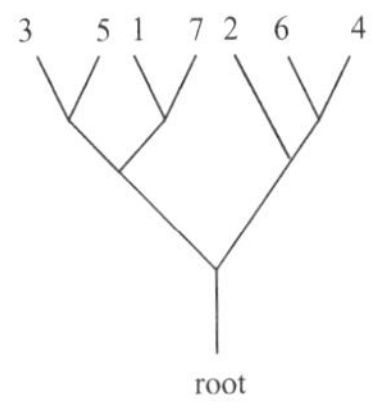

Figure 2.1: A planar binary tree

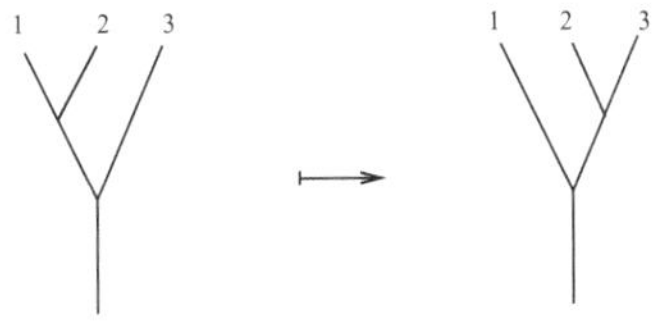

Figure 2.2: An equivalence move

Exercise 1. Show that the operad $\mathcal{T}op(n) = \{$the set of diffeomorphism classes of Riemann spheres with n input holes and 1 output hole$\}$ is isomorphic to the commutative operad of sets.

Exercise 2. Prove that the structure of an algebra over the commutative operad $\mathcal{C}omm$ on a vector space is equivalent to the structure of a commutative associative algebra with a unit.

The associative operad

The *associative operad* $\mathcal{A}ss$ can be considered as a planar one-dimensional analogue of the commutative operad $\mathcal{T}op$. $\mathcal{A}ss(n)$ is the set of equivalence classes of connected planar binary (each vertex being of valence 3) trees that have a root edge and n leaves labeled by integers 1 through n, see Figure 2.1. If $n = 1$, there is only one tree — it has no vertices and only one edge connecting a leaf and a root. If $n = 0$, the only tree is the one with no vertices and no leaves — it only has a root.

Two trees are equivalent if they are related by a sequence of moves of the kind pictured on Figure 2.2, performed over pairs of two adjacent vertices of a tree. The symmetric group acts by relabeling the leaves, as usual. The composition is obtained by grafting the roots of m trees to the leaves of an m-tree, no new vertices being created at the grafting points. Note that this is similar to sewing Riemann surfaces and erasing the seam, just as we did to define operad composition in that case. By definition, grafting a 0-tree to a leaf just removes the leaf and, if this operation creates a vertex of valence 2, we should erase the vertex.

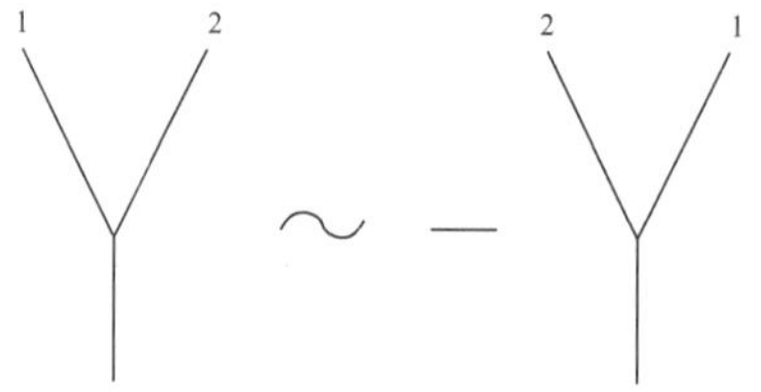

Figure 2.3: Skew symmetry

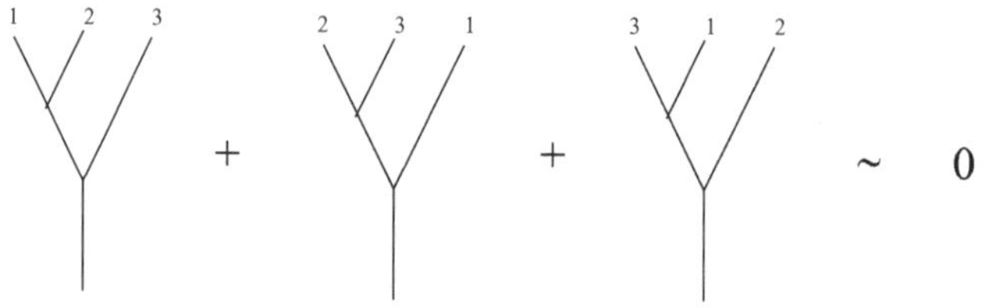

Figure 2.4: The Jacobi identity

Exercise 3. Prove that the structure of an algebra over the associative operad $\mathcal{A}ss$ on a vector space is equivalent to the structure of an associative algebra with a unit.

The Lie operad

The *Lie operad* $\mathcal{L}ie$ is another variation on the theme of a tree operad. Consider the vector space spanned by the same planar binary trees as for the associative operad, except that we do not include a 0-tree, *i.e.*, the operad has only positive components $\mathcal{L}ie(n)$, $n \geq 1$, and there are now two kinds of equivalence relations, see Figures 2.3 and 2.4. Now that we have arithmetic operations in the equivalence relations, we consider the Lie operad as an operad of vector spaces. We also assume that the ground field is of a characteristic other than 2, because otherwise we will arrive at the wrong definition of a Lie algebra.

Exercise 4. Prove that the structure of an algebra over the Lie operad $\mathcal{L}ie$ on a vector space over a field of a characteristic other than 2 is equivalent to the structure of a Lie algebra.

Exercise 5. Describe algebraically an algebra over the operad $\mathcal{L}ie$, if we modify it by including a 0-tree, whose composition with any other tree is defined as (a) zero, (b) the one for the associative operad.

The Poisson operad

Recall that a *Poisson algebra* is a vector space V (over a field of char $\neq 2$) with a unit element e, a dot product ab, and a bracket $[a, b]$ defined, so that the dot

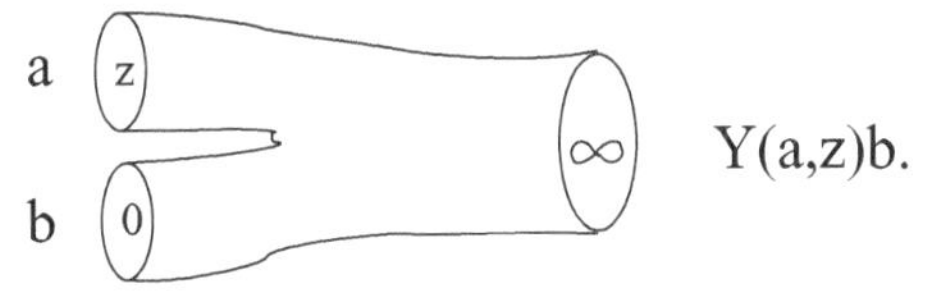

Figure 2.5: The vertex operator

product defines the structure of a commutative associative unital algebra, the bracket defines the structure of a Lie algebra, and the bracket is a derivation of the dot product:

$$[a, bc] = [a, b]c + b[a, c] \qquad \text{for all } a, b, \text{ and } c \in V.$$

Exercise 6. Define the *Poisson operad*, using a tree model similar to the previous examples. Show that an algebra over it is nothing but a Poisson algebra. [*Hint*: Use two kinds of vertices, one for the dot product and the other one for the bracket.]

The Riemann surface operad and vertex operator algebras

Just for a change, let us return to the operad $\mathcal{P}$ of Riemann surfaces, more exactly, isomorphism classes of Riemann spheres with holomorphic holes. What is an algebra over it? Since there are infinitely many nonisomorphic pairs of pants, there are infinitely many (at least) binary operations. In fact, we have an infinite dimensional family of binary operations parameterized by classes of pairs of pants. However modulo the unary operations, those which correspond to cylinders, we have only one fundamental binary operation corresponding to a fixed pair of pants. An algebra over this operad $\mathcal{P}$ is part a CFT data. (For those who understand, this is the tree level, central charge $c = 0$ part). If we consider a holomorphic algebra over this operad, that is, require that the defining mappings $\mathcal{P}(n) \to \mathcal{E}nd_V(n)$, where V is a complex vector space, be holomorphic, then we get part of a chiral CFT, or an object which might have been called a *vertex operator algebra* (*VOA*) in an ideal world. This kind of object is not equivalent to what people use to call a VOA; according to Y.-Z. Huang's theorem, a VOA is a holomorphic algebra over a "partial pseudo-operad of Riemann spheres with rescaling," which is a version of $\mathcal{P}$, where the disks are allowed to overlap. The fundamental operation $Y(a, z)b$ for $a, b \in V$, $z \in \mathbb{C}$ of a VOA is commonly chosen to be the one corresponding to a pair of pants which is the Riemann sphere with a standard holomorphic coordinate and three unit disks around the points 0, z, and ∞ (No doubt, these disks overlap badly, but we shrink them on the figure to look better), see Figure 2.5.

The famous associativity identity

$$Y(a, z - w)Y(b, -w)c = Y(Y(a, z)b, -w)c$$

for vertex operator algebras comes from the natural isomorphism of the Riemann surfaces sketched on Figure 2.6.

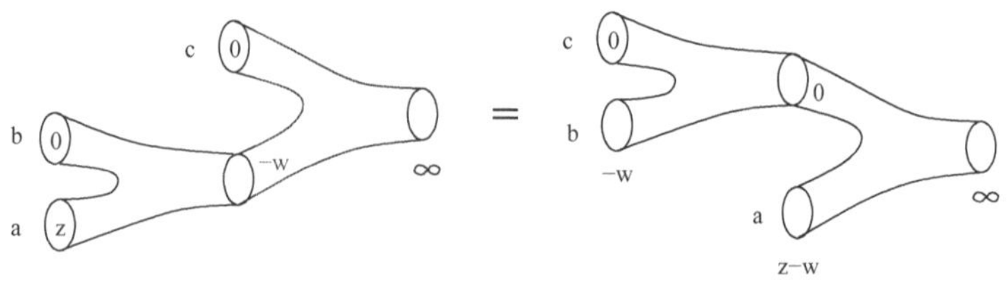

Figure 2.6: VOA associativity

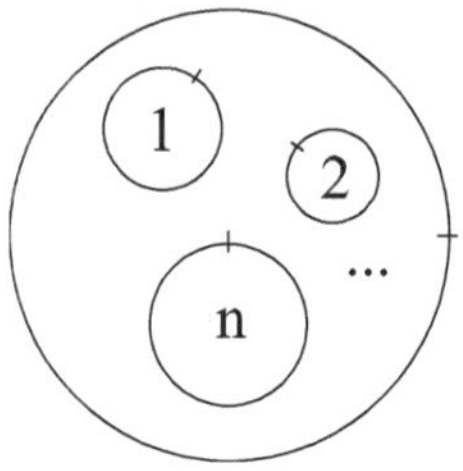

Figure 2.7: Framed little disks

Another remarkable feature of the Riemann surface operad $\mathcal{P}$ is that an algebra over it in the category of spaces group completes to an infinite loop space, which is Tillmann's result [Til97]. This implies, for example, that the classifying space $B\Gamma_\infty^+$ of the stable mapping class group (which is morally the moduli space of Riemann surfaces of infinite genus) is an infinite loop space.

The little disks operads and GBV-algebras

Here is another construction, related to the operad $\mathcal{P}$ of Riemann surfaces and more relevant to string topology. We will be talking about a finite-dimensional retract of $\mathcal{P}$, the *framed little disks operad* $f\mathcal{D}$, which may be defined as follows, see Getzler [Get94] and Markl-Shnider-Stasheff [MSS02]. It is based on the collection $f\mathcal{D} = \{f\mathcal{D}(n) \mid n \geq 1\}$ of configuration spaces of n labeled nonintersecting (closed) disks in the standard (open) unit disk D^2 in the plane $\mathbb{R}^2$ with the choice of a marked point, thought of as framing, on the boundary of each "little" disk, see Figure 2.7. It is also convenient to choose a marked point, e.g., $(1, 0)$, on the boundary of the "big," unit disk, which should not be thought of as extra data, because we are talking about the standard plane $\mathbb{R}^2$, with fixed x and y coordinates. An identity element is the framed little disk coinciding with the big disk, together with framing. The symmetric group acts by relabeling the framed little disks, as usual. The operad composition $\circ_i : f\mathcal{D}(m) \times f\mathcal{D}(n) \to f\mathcal{D}(m+n-1)$ takes a given configuration of n little disks, shrinks it to match the size of the ith little disk in a given configuration of m little disks, rotates the shrunk configuration of n little disks (by a unique element of SO(2)) to match the point on the boundary of the big disk with the point on the boundary of the ith little disk, and glues the configuration of n disks in place of the ith disk, erasing the seam afterwards. A *nonframed version*, which may be defined as the suboperad of $f\mathcal{D}$ with all the points on the boundaries of the little disks point in the direction of the positive x axis, is called the *little disks operad* $\mathcal{D}$.

The little disks operad (along with its higher dimensional version) was invented in topology and proved itself as a powerful tool for studying iterated (based) loop spaces, [May96, BV73]. For example, the little disks operad acts on every based double loop space $\Omega^2 X = \mathrm{Map}_*(S^2, X)$ in the following way. Given a configuration of n little disks and n pointed maps $S^2 \to X$, which we can think of maps $D^2 \to X$ sending the boundary of the standard unit disk D^2 to the basepoint of X, we can define a new map $D^2 \to X$ by using the given maps on the little disks (after appropriate translation and dilation) and extending them to a constant map from the complement of the little disks to X. Of course, the first thing one looks at in topology is homology, and the following description of $H_*(\mathcal{D})$-algebras by F. Cohen is very interesting.

Theorem 2.1.1 (F. Cohen [Coh76]). *An algebra over the homology little disks operad $H_*(\mathcal{D}; k)$ (over a field k of* $\operatorname{char} k \neq 2$*) is equivalent to a* Gerstenhaber *(or simply* G-*)* algebra, *i.e., a graded vector space V with a unit element e, a dot product ab, and a bracket $[a, b]$ defined, so that the dot product defines the structure of a graded commutative associative unital algebra, the bracket defines the structure of a graded Lie algebra on the suspension $V[-1]$, which is the same as V but with a grading shifted by -1, and the bracket is a degree-one derivation of the dot product:*

$$[a, bc] = [a, b]c + (-1)^{(|a|+1)|b|} b[a, c] \qquad \textit{for all } a,\ b, \textit{ and } c \in V.$$

Note that if you have an algebra X over an operad $\mathcal{O}$ in the category of topological spaces, you may always pass to homology and obtain the structure of $H_*(\mathcal{O})$-algebra on the graded vector space $H_*(X)$. Thus, the homology of every double loop space becomes a G-algebra.

Similarly, the framed little disks operad acts on every based double loop space, but the underlying structure of such an action on homology had not been much of interest to topologists till it was discovered in various physical contexts much later under the name of a BV-algebra. A *BV-algebra* is a graded vector space V (over a field of char $\neq 2$) with the structure of a graded commutative algebra and a second-order derivation Δ, called a *BV operator*, of degree one and square zero. The second-order derivation property may be defined using an old idea of Grothendieck:

$$[[[\Delta, L_a], L_b], L_c] = 0 \quad \text{for all } a, b, c \in V,$$

where L_a is the operator of left multiplication by a and the commutators of operators are understood in the graded sense. Alternatively, see Theorem 1.3.1 and [Sch93, Get94], one can define the structure of a BV-algebra on a graded vector space V in the following way:

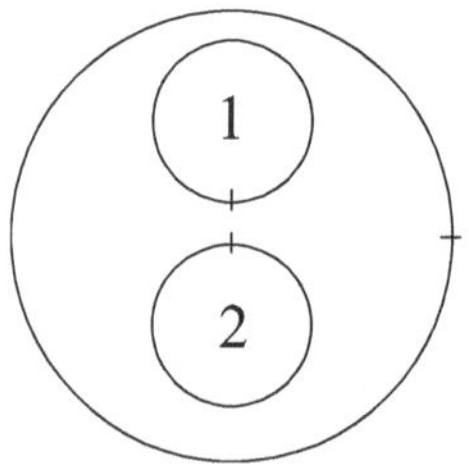

Figure 2.8: The dot product

- A "dot product" $V \otimes V \to V$ and a bracket $V \otimes V \to V[1]$ of degree one on V making it a G-algebra;
- An operator $\Delta : V \to V[1]$ of degree one, which is a differential, i.e., $\Delta^2 = 0$, a degree-one derivation of the bracket, and satisfies the property

$$\Delta(ab) - (\Delta a)b - (-1)^{|a|}a\Delta b = (-1)^{|a|}[a, b].$$

The BV-algebra structure is also known as the algebraic structure induced on homology from the structure of an algebra over the framed little disks operad $f\mathcal{D}^2$ on a topological space:

Theorem 2.1.2 (Getzler [Get94]). *The category of $H_*(f\mathcal{D}; k)$-algebras over a field k of* char $\neq 2$ *is naturally isomorphic to the category of BV-algebras.*

In particular, the homology of a based double loop space $\Omega^2 X$ is naturally a BV-algebra. However, the BV operator and thereby the bracket of this BV-algebra structure are trivial, because the mapping of the equatorial S^1 in S^2 to the double loop space is obviously homotopic to a constant map. This is also confirmed by Salvatore and Wahl's homotopy classification [SW03b] of algebras over the framed little disks operad as double loop spaces $\Omega^2 X$ of S^1-spaces X. The double loop space of an arbitrary space with the natural $f\mathcal{D}$-action is obviously homotopy equivalent to the double loop space with the $f\mathcal{D}$-action coming from the natural $\mathcal{D}$-action on $\Omega^2 X$ and the trivial S^1-action on X.

The proof of Getzler's theorem involves identifying the dot product and the BV operator for a given $H_*(f\mathcal{D}; k)$-algebra V. This is done as follows. Look at the $n = 2$ part

$$H_*(f\mathcal{D}(2); k) \to \mathrm{Hom}(V \otimes V, V)$$

of the operad action. Note that $f\mathcal{D}(2)$ is path connected and take the class of a point in $H_0(f\mathcal{D}(2); k)$, for example, as on Figure 2.8. Define the dot product on V as the resulting bilinear map $V \otimes V \to V$.

To define the BV operator, consider the $n = 1$ part

$$H_*(f\mathcal{D}(1); k) \to \mathrm{Hom}(V, V)$$

of the operad action. Note that $f\mathcal{D}(1)$ is homotopy equivalent to S^1, choose an orientation on S^1, for example, the clockwise rotation of the marked point on the little disk, and take the fundamental class in $H_1(S^1;k) \cong H_1(f\mathcal{D}(1);k)$. The resulting linear map $V \to V$ is the BV operator, by definition.

To complete this proof, one needs to check that the topology of the framed little disks operad forces the right identities between these basic operations, as well as to show that there are no other identities, except those coming from combining the identities in the definition of a BV-algebra. We know how to do that, but we choose not to do it in public.

The little n-disks operad and n-fold loop spaces

Generalizing the little disks operad of the previous section to an arbitrary dimension $n \geq 1$, we get the little n-disks operad (an equivalent version of which is known as the little n-cubes operad) $\mathcal{D}^n$ whose kth component $\mathcal{D}^n(k)$ is the configuration space of k labeled "little" n-dimensional disks inside the standard unit n-disk in $\mathbb{R}^n$. This operad was introduced by Boardman and Vogt [BV73] and May [May96] to recognize n-fold loop spaces $\Omega^n X = \mathrm{Map}_*(S^n, X)$ among other spaces: their recognition principle states that a path-connected topological space is weakly equivalent to an n-fold loop space, if and only if it is weakly equivalent to an algebra over the little n-disks operad.

Passing to homology, F. Cohen also proved that the notion of an algebra over the homology little n-disks operad (over a field of characteristic other than two) is equivalent to the notion of an *n-algebra*, which for $n \geq 2$ is the same as that of a G-algebra, except that the bracket must now have degree $n-1$. A 1-algebra is the same as an associative algebra.

There is also a framed version $f\mathcal{D}^n$ of the little n-disks operad. Here a frame in a little n-disk is a positively oriented orthonormal frame attached to the center of that disk. In other words, each little disk comes with an element of $\mathrm{SO}(n)$. The operad of the rational homology of $f\mathcal{D}^n$ was characterized by the following theorem of Salvatore and Wahl, which will be used in the last chapter on brane topology. For better compatibility with that chapter, we are going to quote this result for $n+1$ rather than n.

Theorem 2.1.3 (Salvatore-Wahl [SW03b]: Theorem 6.5). *The category of algebras over the operad $H_\bullet(f\mathcal{D}^{n+1};\mathbb{Q})$ for $n \geq 1$ is isomorphic to the category of graded vector spaces V over $\mathbb{Q}$ with the following operations and identities. Below $a, b, c \in V$ are homogeneous elements and $|a|$ denotes the degree of a in V.*

1. *A dot product $a \cdot b$, or simply ab, defining the structure of a (graded) commutative associative algebra on V.*
2. *A bracket $[a,b]$ of degree n, defining the structure of a (graded) Lie algebra on the shifted space $V[n]$.*
3. *The bracket with an element a must be a (graded) derivation of the dot product,* i.e., $[a, bc] = [a,b]c + (-1)^{(|a|+n)|b|} b[a,c]$.

4. *For n odd, a collection of unary operators* B_i, $i = 1, \dots, (n-1)/2$, *of degree* $4i - 1$ *and* Δ *of degree* n, *called a* BV operator.
5. *For n even, a collection of unary operators* B_i, $i = 1, \dots, n/2$, *of degree* $4i-1$.
6. *The unary operators* B_i *must square to zero:* $\mathrm{B}_i^2 = 0$ *for all* i. *The operators* B_i *must be (graded) derivations of the commutative algebra structure on* V *and the Lie algebra structure on* $V[n]$.
7. *For n odd,* i.e., *when* Δ *is defined,* $\Delta^2 = 0$, $\Delta(ab) - \Delta(a)b - (-1)^{|a|}a\Delta b = (-1)^{|a|}[a, b]$, *and* $\Delta[a, b] = [\Delta a, b] - (-1)^{|a|}[a, \Delta b]$.

Remark 2. The last identity means that Δ is a graded derivation of the bracket, while the last two equations may be interpreted as Δ being a graded second-order derivation of the dot product.

Idea of proof. This theorem is based on an observation that $f\mathcal{D}^{n+1}$ is a semidirect product $\mathcal{D}^{n+1} \rtimes \mathrm{SO}(n+1)$, see [MSS02]. The elements B_i (and Δ, when it is defined) are the standard generators of $H_\bullet(\mathrm{SO}(n+1); \mathbb{Q})$, Δ corresponding to the Euler class via the transfer map. □

2.1.5 Operads via generators and relations

The tree operads that we looked at above, such as the associative and the Lie operads, are actually operads defined by generators and relations. Here is a way to define such operads in general. To fix notation, assume throughout this section that we work with operads $\mathcal{O}(n)$, $n \geq 1$, of vector spaces.

Definition 2.1.5. An *ideal* in an operad $\mathcal{O}$ is a collection $\mathcal{I}$ of Σ_n-invariant subspaces $\mathcal{I}(n) \subset \mathcal{O}(n)$, for each $n \geq 1$, such that whenever $i \in \mathcal{I}$, its operad composition with anything else is also in $\mathcal{I}$.

The intersection of an arbitrary number of ideals in an operad is also an ideal, and one can define the ideal generated by a subset in $\mathcal{O}$ as the minimal ideal containing the subset.

Definition 2.1.6. For an operad ideal $\mathcal{I} \subset \mathcal{O}$, the *quotient operad* $\mathcal{O}/\mathcal{I}$ is the collection $\mathcal{O}(n)/\mathcal{I}(n)$, $n \geq 1$, with the structure of operad induced by that on $\mathcal{O}$.

The *free operad* $F(S)$ *generated by a collection* $S = \{S(n) \mid n \geq 1\}$ *of sets*, is defined as follows.

$$F(S)(n) = \bigoplus_{n\text{-trees } T} k \cdot S(T),$$

where the summation runs over all planar rooted trees T with n labeled leaves and

$$S(T) = \mathrm{Map}(v(T), S),$$

the set of maps from the set $v(T)$ of vertices of the tree T to the collection S assigning to a vertex v with $\mathrm{In}(v)$ incoming edges an element of $S(\mathrm{In}(v))$ (the

Figure 2.9: The operad identity

edges are directed toward the root). In other words, an element of $F(S)(n)$ is a linear combination of planar n-trees whose vertices are decorated with elements of S. There is a special tree with no vertices, see Figure 2.9. The component $F(S)(1)$ contains, apart from $S(1)$, the one-dimensional subspace spanned by this tree.

The following data defines an operad structure on $F(S)$.

1. The identity element is the special tree in $F(S)(1)$ with no vertices.
2. The symmetric group Σ_n acts on $F(S)(n)$ by relabeling the inputs.
3. The operad composition is given by grafting the roots of trees to the leaves of another tree. No new vertices are created.

Definition 2.1.7. Now let R be a subset of $F(S)$, *i.e.*, a collection of subsets $R(n) \subset F(S)(n)$. Let (R) be the ideal in $F(S)$ generated by R. The quotient operad $F(S)/(R)$ is called the *operad with generators S and defining relations R*.

Example 2.1.6. The associative operad $\mathcal{A}ss$ is the operad generated by a point $S = S(2) = \{\bullet\}$ with a defining relation given by the associativity condition, see Section 2.1.4, expressed in terms of trees. Note that equation $S = S(2)$ implies that $S(n) = \varnothing$ for $n \neq 2$.

Example 2.1.7. The Lie operad $\mathcal{L}ie$ is the operad also generated by a point $S = S(2) = \{\bullet\}$ with defining relations given by the skew symmetry and the Jacobi identity, see Section 2.1.4.

Example 2.1.8. The Poisson operad is the operad generated by a two-point set $S = S(2) = \{\bullet, \circ\}$ with defining relations given by the commutativity and the associativity for simple trees decorated only with $\bullet$'s, the skew symmetry and the Jacobi identity for simple trees decorated with $\circ$'s, and the Leibniz identity for binary 3-trees with mixed decorations, see Section 2.1.4.

2.2 The cacti operad

The construction and results in this section have been announced in [Vor01]. The BV structure arising in string topology at the level of homology comes from an action of a *cacti operad* $\mathcal{C}$ at the motivic level, quite close to the category of topological spaces.

The kth component $\mathcal{C}(k)$ of the cacti operad $\mathcal{C}$ for $k \geq 1$ may be described as follows. $\mathcal{C}(k)$ is the set of tree-like configurations of parameterized circles, called the

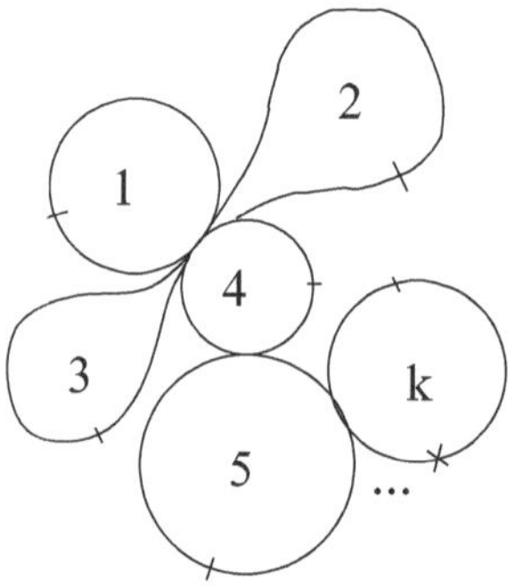

Figure 2.10: A cactus

lobes, labeled by numbers 1 through k, of varying (positive) radii, along with the following data: (1) the choice of a cyclic order of components at each intersection point and (2) the choice of a marked point on the whole configuration along with the choice of one of the circles on which this point lies. The last choice is essential only when the marked point happens to be an intersection point. Here "tree-like" means that the dual graph of this configuration, whose vertices correspond to the lobes and the intersection points thereof and whose edges reflect the obvious incidence relation, is a tree. We will refer to an element of $\mathcal{C}(k)$ as a *cactus*. A cactus defines a *pinching map* from the standard unit circle S^1 to the cactus. The pinching map starts from the marked point in the direction of the increasing parameter on the circle on which the marked point lies and traces the whole cactus along the parameters on the circles, jumping from one lobe to the next in the cyclic order at the intersection points. This produces a map from the circle of circumference c equal to the total circumference of the cactus. To get a map from the standard unit circle, first expand (or contract) it to a circle of circumference c.

The topology on the set $\mathcal{C}(k)$ of cacti may be introduced in the following way. There is a unique way up to isotopy to place a cactus on the plane, so that the parameters of the lobes go counterclockwise and the cyclic order of the lobes at each intersection point is also counterclockwise, see Figure 2.10. Thereby a cactus defines a metric planar graph whose vertices are the intersection points of the circles and edges are the arcs between two adjacent vertices. The word "metric" refers to the fact that the edges are provided with positive real numbers, the arclengths in the parameters of the lobes. A cactus is determined by its metric planar graph and the choice of a marked point on each of the k circles forming the cactus (called "interior boundary components" of the graph) and a global marked point on the "exterior boundary component" defined as the standard circle S^1 together with the pinching map.

First of all, we will define a topology on the space of metric planar graphs (and thereby on its subspace of metric planar graphs arising from cacti) and then describe the space $\mathcal{C}(k)$ of cacti as an $(S^1)^{k+1}$ fiber bundle over the subspace of metric planar graphs.

Each planar graph Γ defines an open cell $\mathbb{R}_+^{e(\Gamma)}$, where $\mathbb{R}_+$ is the interval $(0, \infty)$ of the real line and $e(\Gamma)$ is the set of edges of Γ. This open cell is attached to the union of lower dimensional cells as follows. Consider part of the boundary of $\mathbb{R}_+^{e(\Gamma)}$ in $\mathbb{R}^{e(\Gamma)}$ given by setting some of the edge lengths to zero, except for the edges forming simple loops, i.e., the edges which span the whole lobe less a point. An attaching map identifies the face $l(e_0) = 0$ of the cell $\mathbb{R}_+^{e(\Gamma)} = \{l(e) > 0 \mid e \in e(\Gamma)\}$ with a cell $\mathbb{R}_+^{e(\Gamma/e_0)}$, where Γ/e_0 is the planar graph obtained from Γ by contracting the edge e_0 to a point. This way we build the space of metric planar graphs as a generalization of a cell complex: start with a collection of open cubes of dimension k (corresponding to the planar wedges of k circles whose starting points and the global marked point are at the basepoint of the wedge), attach a collection of open cubes of dimension $k+1$ to them along some of their faces, then attach open cubes of dimension $k+2$ to the result, and so on. We take the topology of the union on the resulting space. This topologizes the space of metric planar graphs. The "universal graph bundle" over this space is determined by specifying the fiber over a point to be the graph (thought of as a one-dimensional CW complex) represented by this point in the space of graphs. Marking a point on a specific (interior or exterior) boundary component defines an S^1-bundle over the space of graphs. Thus, $\mathcal{C}(k)$ becomes a product of $k+1$ such bundles over the corresponding subspace of metric planar graphs.

Remark 3. This construction identifies the space $\mathcal{C}(k)$ of cacti as a certain subspace of the space of metric ribbon graphs of genus zero with $k+1$ labeled boundary components and one point marked on each boundary component. See Chapter 3 for a discussion of general ribbon (fat) graphs.

The *operad structure* on the cacti comes from the pinching map. Given two cacti and the ith lobe in the first one, the operad composition $\circ_i$ will be given by further gluing the ith lobe of the first cactus (identified with the standard circle S^1 via a suitable dilation) to the second cactus along the pinching map from S^1 to the second cactus.

Theorem 2.2.1 ([Vor01]). *The cacti operad $\mathcal{C}$ is homotopy equivalent to the framed little disks operad $f\mathcal{D}^2$.*

Remark 4. In principle, one can prove this theorem explicitly, as indicated in [CJ02]. For example, P. Salvatore (private communication) suggests constructing a map from the configuration space of k labeled points in the plane by placing at each point a particle, which creates a radial repulsive field of magnitude $1/r$, where r is the distance from the particle, and looking at the degenerate trajectories of the superposition field. This is, in fact, dual to the construction of a ribbon graph on the Riemann sphere with $k+1$ punctures using a Strebel differential and its degenerate horizontal trajectories: Salvatore's approach uses the vertical ones.

We will take a less constructive approach and deduce the statement from Salvatore-Wahl's recognition principle [SW03b] for the framed little disks operad

$f\mathcal{D}^2$, which generalizes Fiedorowicz's recognition principle [Fie92, Fie98] for the little disks operad $\mathcal{D}^2$. We will use the recognition principle in the following form. Let B_k be the braid group on k strands, $k \geq 1$, then the ribbon braid group RB_k on k ribbons is a semidirect product $\mathbb{Z}^k \rtimes B_k$ determined by the permutation action of B_k on $\mathbb{Z}^k$. It is worth pointing out that the ribbon braid group RB_k is a subgroup of B_{2k} by the induced braid on the edges of the ribbons. Let PRB_k denote the kernel of the natural epimorphism $\mathrm{RB}_k \to \Sigma_k$. Then $\mathrm{PRB}_k \cong \mathbb{Z}^k \times \mathrm{PB}_k$, where PB_k is the pure braid group, the kernel of the natural epimorphism $B_k \to \Sigma_k$.

Theorem 2.2.2 (Salvatore-Wahl [SW03b]). *Let $\mathcal{O} = \{\mathcal{O}(k) \mid k \geq 1\}$ be a topological operad, satisfying the following conditions:*

1. *Each quotient $\mathcal{O}(k)/\Sigma_k$ is $K(\mathrm{RB}_k, 1)$, the normal subgroup $\mathrm{PRB}_k \subset \mathrm{RB}_k$ corresponding to the covering $\mathcal{O}(k) \to \mathcal{O}(k)/\Sigma_k$.*
2. *There exists an operad morphism $\mathcal{D}^1 \to \mathcal{O}$ from the little intervals operad $\mathcal{D}^1$.*

Then the operad $\mathcal{O}$ is homotopy equivalent to the framed little disks operad $f\mathcal{D}^2$.

Remark 5. If we require in Condition 2 that a cofibrant model of the little intervals operad $\mathcal{D}^1$ admits a morphism to $\mathcal{O}$, the two conditions become not only sufficient, but also necessary. Examples of cofibrant models of $\mathcal{D}^1$ include the operad of metrized planar trees with labeled leaves (a cellular operad, whose chain operad is A_∞) and Boardman-Vogt's W-resolution $W\mathcal{D}^1$ of $\mathcal{D}^1$.

Remark 6. This recognition principle can be generalized to any non-Σ operad of groups $H(k)$ with $H(1)$ abelian and the operad composition and the operad unit for $H(1)$ given by the group law and the group unit therein, respectively. This operad of groups should also be provided with an operad epimorphism $H(k) \to \Sigma_k$. $\{\Sigma_\bullet\}$, $\{B_\bullet\}$, and $\{\mathrm{RB}_\bullet\}$ are examples of such operads of groups. The proof of the recognition principle is the same.

Proof. We will only sketch a proof, following Salvatore and Wahl: the reader may collect missing details from [Fie92, Fie98, MS02, SW03b]. One first introduces the notion of a ribbon braided operad, in which the symmetric groups get replaced with the ribbon braid groups. Condition 2 amounts to a consistent choice of basepoints in the universal coverings $\widetilde{\mathcal{O}}(k)$, which gives the structure of a ribbon braided operad on $\widetilde{\mathcal{O}}$. Condition 1 implies that each space $\widetilde{\mathcal{O}}(k)$ is contractible with a free action of RB_k. Then the following diagram of homotopy equivalences of ribbon braided operads

$$\widetilde{f\mathcal{D}^2} \leftarrow \widetilde{f\mathcal{D}^2} \times \widetilde{\mathcal{O}} \to \widetilde{\mathcal{O}},$$

after taking quotients by the pure ribbon braid groups PRB_k, gives homotopy equivalences of usual operads:

$$f\mathcal{D}^2 \leftarrow (\widetilde{f\mathcal{D}^2} \times \widetilde{\mathcal{O}})/\,\mathrm{PRB}_\bullet \to \mathcal{O}. \qquad \square$$

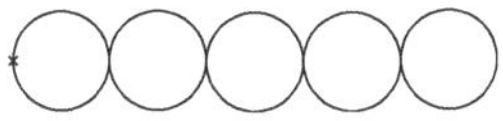

Figure 2.11: The basepoint in the cacti operad

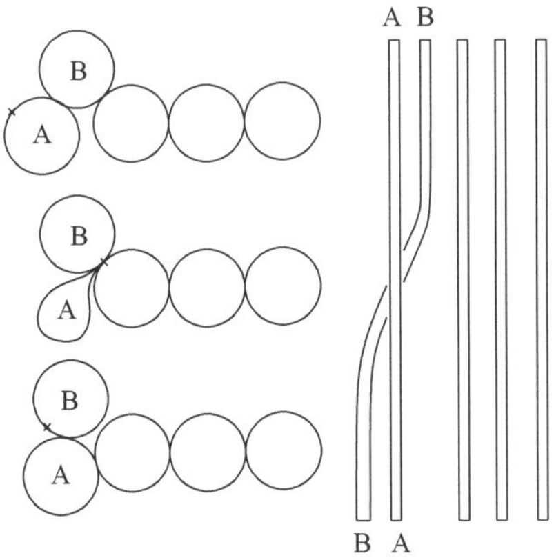

Figure 2.12: A path in the cacti space corresponding to a braid

Proof of Theorem 2.2.1. As promised, we will use Salvatore-Wahl's recognition principle, Theorem 2.2.2.

To see that Σ_k acts freely on $\mathcal{C}(k)$, note that the structure map from S^1 to the cactus defines an ordering on the lobes.

There is an obvious homomorphism $\mathrm{RB}_k \to \pi_1(\mathcal{C}(k)/\Sigma_k)$ in which a ribbon braid moves in the plane the lobes of a fixed cactus, *e.g.*, one like on Figure 2.11, keeping the marked point somewhere on the left to return it to the original position at the end of the move. Here is an example of three intermediate snapshots of the above cactus through a move, with the corresponding braid on the side, see Figure 2.12.

We will prove that $\mathcal{C}(k)/\Sigma_k$ is $K(\mathrm{RB}_k, 1)$ by induction on k. For $k = 1$ the space $\mathcal{C}(1)/\Sigma_1 = \mathcal{C}(1)$ is obviously homeomorphic to $S^1 \times \mathbb{R}_+$, which is $K(\mathbb{Z}, 1)$. Since $\mathrm{RB}_1 \cong \mathbb{Z}$, we get the induction base.

To make the induction step, assume that the above homomorphism $\mathrm{RB}_k \to \pi_1(\mathcal{C}(k)/\Sigma_k)$ is bijective and consider a forgetful fibration $\mathcal{C}(k+1) \to \mathcal{C}(k)$ which contracts the $k+1$st lobe to a point, *i.e.*, takes a quotient of the cactus by its $k+1$st lobe. The marked point, if it happened to be on that lobe, will be at the resulting contraction point, assigned to the lobe that follows the one being contracted, if one moves from the marked point along the parameter of the wrapping map from S^1 to the cactus. It is easy to observe that the fiber of this fibration is homotopy equivalent to $\left(\bigvee_n S^1\right) \times S^1$, where the last circle is given by the angular parameter on the contracted circle.

This fibration admits a section, for example, by placing a unit circle at the marked point, so that the wrapping map will trace this circle completely from its angular parameter 0 and at 2π bumps into the marked point.

Therefore, the long exact homotopy sequence of this fibration splits into short exact sequences as follows:

$$1 \to \pi_1\left(\left(\bigvee_k S^1\right) \times S^1\right) \to \pi_1(\mathcal{C}(k+1)) \to \pi_1(\mathcal{C}(k)) \to 1,$$
$$0 \to 0 \to \pi_i(\mathcal{C}(k+1)) \to \pi_i(\mathcal{C}(k)) \to 0 \qquad \text{for } i \geq 2,$$

with the arrow $\pi_1(\mathcal{C}(k+1)) \to \pi_1(\mathcal{C}(k))$ in the first sequence having a right inverse. By induction, $\mathcal{C}(k)$ is $K(\mathrm{PRB}_k, 1)$. Then from the short exact sequences, we see that $\pi_i(\mathcal{C}(k+1)) = 0$ for $i \geq 2$.

To treat the fundamental-group case, note that we have a natural morphism of group extensions:

$$\begin{array}{ccccccccc} 1 & \longrightarrow & \pi_1\left(\left(\bigvee_k S^1\right) \times S^1\right) & \longrightarrow & \mathrm{PRB}_{k+1} & \longrightarrow & \mathrm{PRB}_k & \longrightarrow & 1 \\ & & \| & & \downarrow & & \downarrow & & \\ 1 & \longrightarrow & \pi_1\left(\left(\bigvee_k S^1\right) \times S^1\right) & \longrightarrow & \pi_1(\mathcal{C}(k+1)) & \longrightarrow & \pi_1(\mathcal{C}(k)) & \longrightarrow & 1, \end{array}$$

where the first line is obtained by applying the fundamental group functor to a similar forgetful fibration $f\mathcal{D}(k+1) \to f\mathcal{D}(k)$. The right vertical arrow is an isomorphism by the induction assumption, therefore, so is the middle one.

Now the following morphism of extensions:

$$\begin{array}{ccccccccc} 1 & \longrightarrow & \mathrm{PRB}_{k+1} & \longrightarrow & \mathrm{RB}_{k+1} & \longrightarrow & \Sigma_{k+1} & \longrightarrow & 1 \\ & & \downarrow & & \downarrow & & \| & & \\ 1 & \longrightarrow & \pi_1(\mathcal{C}(k+1)) & \longrightarrow & \pi_1(\mathcal{C}(k+1)/\Sigma_{k+1}) & \longrightarrow & \Sigma_{k+1} & \longrightarrow & 1 \end{array}$$

shows that the middle vertical arrow is an isomorphism. This completes the induction step and thereby verifies Condition 1.

Finally, let us check Condition 2. Given a configuration of k little intervals, sketch k circles based on the little intervals as diameters. Then contract the spaces between the circles and put the marked point at the left corner of the first circle on the left, as on Figure 2.13. Amazingly enough, this defines a morphism of operads $\mathcal{D}^1 \to \mathcal{C}$. □

2.3 The cacti action on the loop space

Let M be an oriented smooth manifold of dimension d. We would like to study continuous k-ary operations on the *free loop space* $LM := M^{S^1} := \mathrm{Map}(S^1, M)$ of continuous loops, functorial with respect to M. By passing to singular chains or homology, these operations will induce functorial operations on the chain and homology level, respectively. We will generalize a homotopy-theoretic approach developed by R. Cohen and J. D. S. Jones [CJ02] for the loop product.

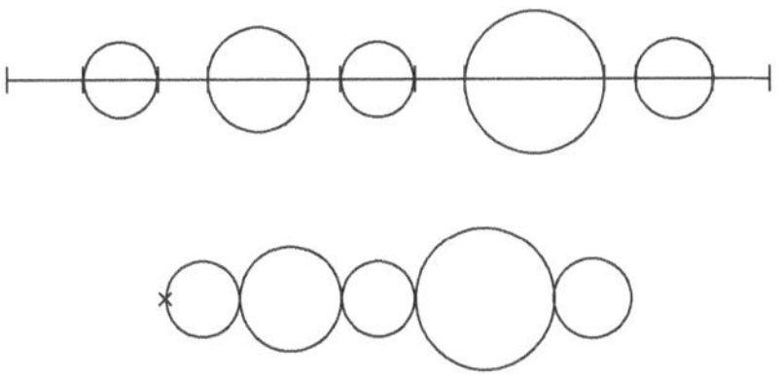

Figure 2.13: Little intervals to cacti

2.3.1 Action via correspondences

First of all, consider the following diagram

$$\mathcal{C}(k) \times (LM)^k \xleftarrow{\rho_{\text{in}}} \mathcal{C}(k)M \xrightarrow{\rho_{\text{out}}} LM, \tag{2.1}$$

where $\mathcal{C}$ is the cacti operad, $k \geq 1$, and $\mathcal{C}(k)M$ is the space of pairs (c, f) with $c \in \mathcal{C}(k)$ being a cactus and $f : c \to M$ a continuous map of the cactus c to M. The map $\rho_{\text{in}} : \mathcal{C}(k)M \to \mathcal{C}(k) \times (LM)^k$ takes a pair (c, f) to $c \in \mathcal{C}(k)$ and the restrictions of f to the k lobes of c. It is an embedding of codimension $d(k-1)$. The map $\rho_{\text{out}} : \mathcal{C}(k)M \to LM$ composes the pinching map $S^1 \to c$ with $f : c \to M$. This diagram is, in principle, all you need to define an operad "action", *i.e.*, the structure of a "$\mathcal{C}(k)$-algebra" on the loop space LM. It would be an honest operad action, should the arrow ρ_{in} be pointing in the opposite direction and the resulting composite map be compatible with the operad structure on $\mathcal{C}$. To check this compatibility, we indeed need to find a way to invert ρ_{in} and take the composite map. The main idea is to take the motivic standpoint and treat diagram (2.1) as a single morphism going from left to right.

We will consider the *category Corr of correspondences.* The objects of this category are just topological spaces, and a morphism between two objects X and Y is a *correspondence*, by which here we mean a diagram $X \leftarrow X' \to Y$ of continuous maps for some space X', or, equivalently, a map $X' \to X \times Y$. One composes two correspondences $X \leftarrow X' \to Y$ and $Y \leftarrow Y' \to Z$ by taking a pullback

$$\begin{array}{ccc} X' \times_Y Y' & \longrightarrow & Y' \\ \downarrow & & \downarrow \\ X' & \longrightarrow & Y \end{array}$$

which defines a new correspondence $X \leftarrow X' \times_Y Y' \to Z$.

Theorem 2.3.1. 1. *Diagram (2.1), considered as a morphism $\mathcal{C}(k) \times (LM)^k \to LM$ in $\mathcal{C}orr$, defines the structure of a $\mathcal{C}$-algebra on the loop space LM in $\mathcal{C}orr$.*

2. *This $\mathcal{C}$-algebra structure on the loop space LM in $\mathcal{C}orr$ induces an $h_*(\mathcal{C})$-algebra structure on the shifted homology $h_{*+d}(LM)$ for any multiplicative generalized homology theory h_* which supports an orientation of M.*

Proof. 1. We just need to see that the following diagram commutes in $\mathcal{C}orr$:

$$\begin{array}{ccc} \mathcal{C}(k) \times \mathcal{C}(l) \times (LM)^{k+l-1} & \xrightarrow{\circ_i \times \mathrm{id}} & \mathcal{C}(k+l-1) \times (LM)^{k+l-1} \\ \downarrow & & \downarrow \\ \mathcal{C}(k) \times (LM)^k & \longrightarrow & LM, \end{array} \tag{2.2}$$

where the unmarked arrows are correspondences defined using diagram (2.1), the left vertical arrow is such a correspondence preceded by an appropriate permutation, to make sure that $\mathcal{C}(l)$ acts on the l components LM in LM^{k+l-1}, starting from the ith one. To verify the commutativity of diagram (2.2), we compose the correspondences, using pullbacks, and see that both compositions (the lower left one and the upper right one) are equal to the following correspondence:

$$\mathcal{C}(k) \times \mathcal{C}(l) \times (LM)^{k+l-1} \leftarrow \mathcal{C}(k) \circ_i \mathcal{C}(l)M \to LM,$$

where $\mathcal{C}(k) \circ_i \mathcal{C}(l)M := \{(C_1, C_2, f) \mid C_1 \in \mathcal{C}(k), C_2 \in \mathcal{C}(l)$, and $f : C_1 \circ_i C_2 \to M$ is continuous$\}$ and the maps are obvious.

2. Verifying the statement at the homology level is a little subtler, as not any correspondence induces a morphism on homology. To verify we have such a morphism in this case, we follow the approach of Cohen and Jones [CJ02] as described in Chapter 1. Suppose we have a correspondence $X \xleftarrow{e} X' \xrightarrow{\gamma} Y$ between smooth (infinite dimensional) manifolds, so that e is a regular embedding of codimension p. (One can say that such a *correspondence is of degree* $-p$.) A *regular embedding* $X \hookrightarrow X'$ is, locally in X' at X, the product of a neighborhood in X with a Euclidean space of codimension p. This condition assures that the tubular neighborhood theorem applies. Then we can apply the Thom collapse construction to e and get a composition

$$h_*(X) \xrightarrow{e^!} h_{*-p}(X') \xrightarrow{\gamma_*} h_{*-p}(Y).$$

Note that the inclusions $\mathcal{C}(k) \circ_i \mathcal{C}(l)M \hookrightarrow \mathcal{C}(k) \times \mathcal{C}(l)M \times LM^{k-1} \hookrightarrow \mathcal{C}(k) \times \mathcal{C}(l) \times (LM)^{k+l-1}$ participating in the lower left path on diagram (2.2) are regular embeddings of codimensions $d(k-1)$ and $d(l-1)$, respectively, because they are pullbacks of regular (finite-dimensional) embeddings along fiber bundles, while the composite inclusion, which also shows up in the upper right path on (2.2), is a regular embedding of codimension $d(k+l-2)$. Because of the functoriality of the homology with respect to Thom collapse maps and the naturality of Thom collapse maps on pullbacks of a regular embedding along a transversal regular immersion, we conclude that diagram (2.2) induces a commutative diagram in homology. This checks that $h_{*+d}(LM)$ is an algebra over the operad $h_*(\mathcal{C})$. □

2.3.2 The BV structure

String topology originated from Chas and Sullivan's construction of a BV structure on the shifted homology of a loop space in a compact, oriented manifold. A BV-algebra is nothing but another algebraic structure, see Section 2.1.4. However, its constant appearance within different contexts of mathematical physics makes it worth a little attention.

Combining Theorems 2.2.1 and 2.3.1 with Theorem 2.1.2 and checking what the basic operations (the dot product and the BV operator) really are, as in the discussion of the proof of Theorem 2.1.2, we conclude with the following result.

Corollary 2.3.2. *For an oriented d-manifold M and a field k of characteristic* $\neq 2$, *the space* $\mathbb{H}_*(LM;k) = H_{*+d}(LM;k)$ *has the natural structure of a BV-algebra. This structure coincides with the one constructed by Chas and Sullivan.*

Chapter 3

Field theoretic properties of string topology

3.1 Field theories

3.1.1 Topological Field Theories

The first axiomatic definition of Topological (Quantum) Field Theories was due to Atiyah [Ati88] and based on the work of Witten and Segal. An *n-dimensional Topological Field Theory* (*TFT*) comprises the following data.[1]

1. An assignment: {a closed oriented $(n-1)$-dimensional manifold X, considered up to diffeomorphism} $\mapsto$ {a complex vector space $V(X)$}, a *state space*, interpreted as the space of functions (or other linearizations of a space, such as differential forms, (singular) cochains, or chains) on a space of fields associated to X. In the case of the so-called sigma-model, most relevant to string topology, the space of fields is the loop space LX.

2. An assignment: {an n-dimensional oriented cobordism Y, considered up to diffeomorphism, bounded by two $(n-1)$-manifolds X_1 and X_2} $\mapsto$ {a linear operator $\Psi_Y : V(X_1) \to V(X_2)$}. One usually thinks of X_1 as the *input* and X_2 as the *output* of cobordism Y.

These data need to satisfy the following axioms:

$$V(X_1 \coprod X_2) \xrightarrow{\sim} V(X_1) \otimes V(X_2)$$

and Ψ_Y must be compatible with the tensor products and composition of cobordisms.

[1] One usually puts in more data, so what we are describing is a geometric background of a TFT.

Remark 7. The categories of cobordisms and vector spaces are tensor (more precisely, symmetric monoidal) categories and an n-dimensional TFT is nothing but a tensor functor between them.

Remark 8. For $n = 2$, a closed oriented one-dimensional manifold X is diffeomorphic to a disjoint union $\coprod S^1$ of finitely many circles S^1, interpreted as closed strings. In this case, the category of cobordisms is a PROP, see Section 2.1, and a TFT is a morphism from the PROP of compact oriented surfaces with boundary to $\mathcal{E}nd_V$, the endomorphism PROP of the vector space V. The PROP $\mathcal{E}nd_V$ may be identified with the tensor subcategory of the category of vector spaces generated by V.

Theorem 3.1.1 (Folklore). *A 2-dimensional TFT is equivalent to a complex Frobenius algebra A, which is a finite-dimensional commutative associative algebra A with a unit and a nonsingular trace $\theta : A \to \mathbb{C}$, i.e., a linear map defining a nonsingular bilinear product $A \otimes A \to \mathbb{C}$ via $\theta(ab)$, cf. Section* 1.1.

Remark 9. This is a folk theorem, whose proof was passed on since the early nineties by word of mouth by many folks, at least including Dijkgraaf, Segal, and Witten, and was written down in the mathematical literature by Abrams [Abr96] and Segal [Seg99].

Outline of Proof. Let V be the state space of a TFT. We would like to construct the structure of a Frobenius algebra on V. Consider the oriented surfaces and the corresponding operators, which we denote ab, $\langle a, b\rangle$, *etc.*, as on Figure 3.1.

We claim that these operators define the structure of a Frobenius algebra on V. Indeed, the multiplication is commutative, because if we interchange labels at the legs of a pair of pants we will get a homeomorphic oriented surface. Therefore, the corresponding operator $a \otimes b \mapsto ba$ will be equal to ab. Similarly, the associativity $(ab)c = a(bc)$ of multiplication is based on the fact that the two surfaces on Figure 3.2 are homeomorphic: The property $ae = ea = a$ of the unit element comes from the diffeomorphism on Figure 3.3. Thus, we see that V is a commutative associative unital algebra. Now, the diffeomorphism on Figure 3.4 proves the identity $\langle a, b\rangle = \operatorname{tr}(ab)$, which, along with the associativity, implies $\langle ab, c\rangle = \langle a, bc\rangle$. The fact that the inner product $\langle\, ,\rangle$ is nondegenerate follows from the diffeomorphism on Figure 3.5, which implies that the composite mapping

$$\begin{array}{ccccc} V & \xrightarrow{\operatorname{id}\otimes\psi} & V\otimes V\otimes V & \xrightarrow{\langle,\rangle\otimes\operatorname{id}} & V, \\ v & \longmapsto & \sum_{i=1}^n v\otimes u_i\otimes v_i & \longmapsto & \sum_{i=1}^n \langle v, u_i\rangle v_i \end{array}$$

is equal to $\operatorname{id} : V \to V$. It takes a little linear-algebra exercise to see that this implies that $\dim V < \infty$ and the symmetric bilinear form is nondegenerate. This completes the construction of the structure of a finite-dimensional Frobenius algebra on the state space V of a TFT.

Conversely, if we have a finite-dimensional Frobenius algebra V, we can define the structure of a TFT on the vector space V by (1) cutting a surface down into

Figure 3.1: Basic operations corresponding to surfaces

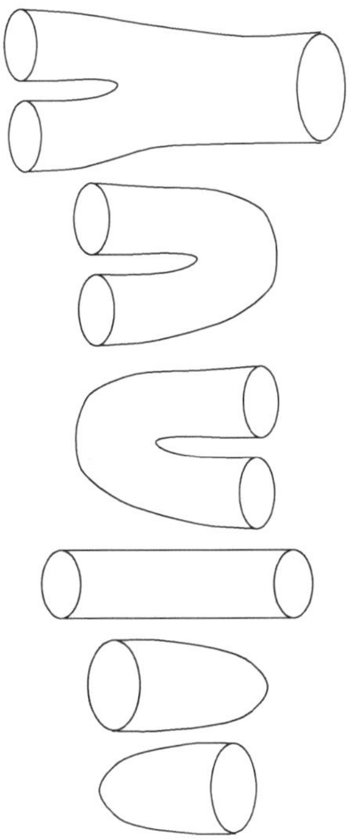

$\longmapsto \quad V \otimes V \to V, \quad a \otimes b \mapsto ab$

$\longmapsto \quad V \otimes V \to k, \quad a \otimes b \mapsto \langle a, b \rangle$

$\longmapsto \quad k \to V \otimes V, \quad 1 \mapsto \psi$

$\longmapsto \quad \mathrm{id} : V \to V$

$\longmapsto \quad \mathrm{tr} : V \to k$

$\longmapsto \quad k \to V, \quad 1 \mapsto e$

Figure 3.2: Associativity

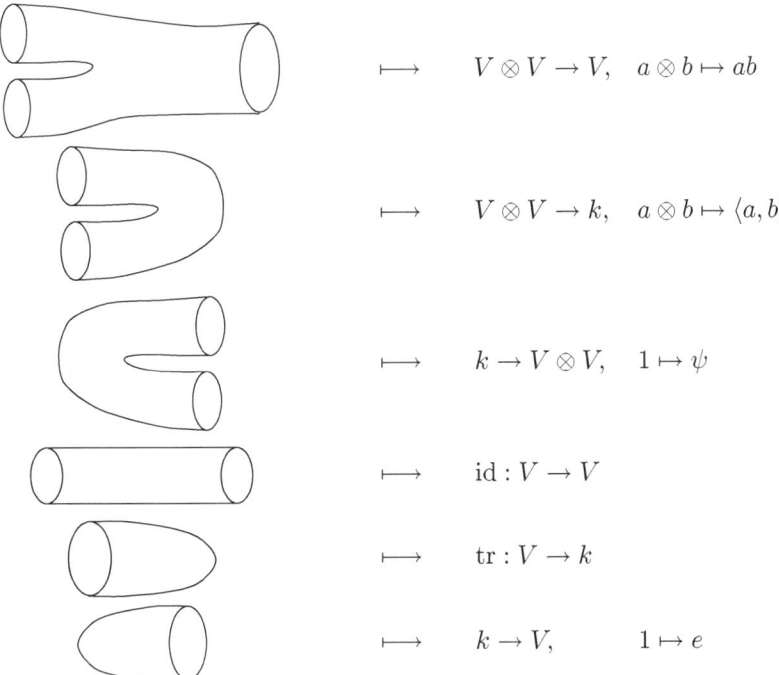

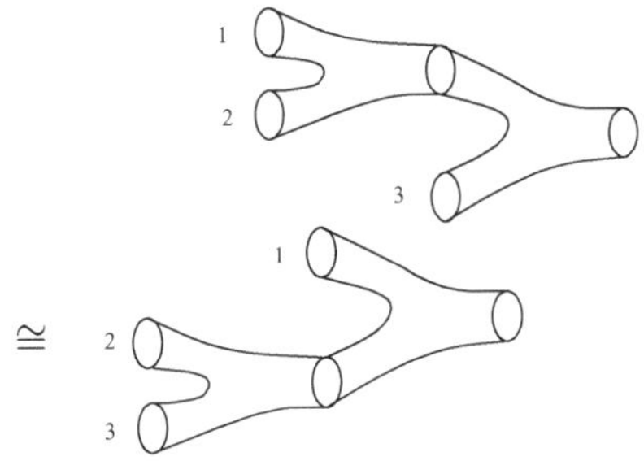

Figure 3.3: The unit axiom

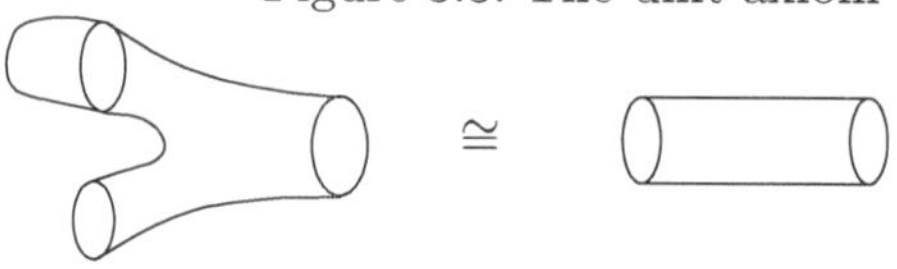

Figure 3.4: The trace

pairs of pants, cylinders, and caps; (2) defining the operators corresponding to those basic objects using the multiplication (or its linear dual), the identity map, and the unit element $e \in V$ (or the dual of the map $k \to V$, $1 \mapsto e$, as the trace functional), respectively; and (3) using the sewing axiom. The fact that the composite operator is independent of the way we cut down the surface follows from Figure 3.2 and the associativity of multiplication. □

Example 3.1.1 (The Toy Model, Dijkgraaf and Witten [DW90]). Let G be a finite group. Assign to any closed one-manifold $X = \coprod S^1$ the space $V(X)$ of $\mathbb{C}$-valued functions on the isomorphism classes of principal G-bundles over X. The set of these isomorphism classes may be identified with $\mathrm{Hom}(\pi_1(X), G)/G$, the set of conjugacy classes of homomorphisms $\pi_1(X) \to G$. To an oriented surface Y with no inputs, assign the operator $\Psi_Y : \mathbb{C} \to V(\partial Y)$ defined by the equation

$$\Psi_Y(P) := \sum_{Q:Q|_{\partial Y}=P} \frac{1}{|\mathrm{Aut}\, Q|}$$

as a function on the set of principal G bundles P over ∂Y. In this case, $A = V(S^1)$ becomes the center of the group algebra $\mathbb{C}[G]$ with Frobenius' trace $\theta(\sum_g \lambda_g g) = \lambda_e/|G|$, where $e \in G$ is the unit element. Note that the same construction yields an n-dimensional TFT for any $n \geq 1$.

Figure 3.5: Nondegeneracy

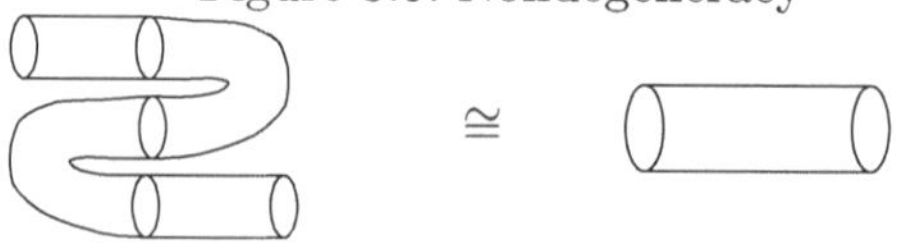

3.1.2 (Topological) Conformal Field Theories

Let us concentrate on the case $n = 2$ and, instead of the very small PROP of oriented smooth surfaces, consider the Segal PROP $\mathcal{P}$ of Riemann surfaces (i.e., complex curves) with holomorphic holes from Example 2.1.2. Here we will be discussing what is known as QFTs with zero central charge.

Recall that in Chapter 2 we introduced the following notion.

Definition 3.1.1. A *Conformal Field Theory (CFT)* is an algebra over the PROP $\mathcal{P}$.

We would like to consider variations on this theme.

Definition 3.1.2. A *Cohomological Field Theory-I (CohFT-I)* is an algebra over the homology PROP $H_*(\mathcal{P})$. The Roman numeral one in the name is to distinguish this theory from a standard *Cohomological Field Theory (CohFT)*, which is an algebra over the PROP $H_*(\overline{\mathcal{M}})$, where $\overline{\mathcal{M}}$ is the PROP of moduli space of stable compact algebraic curves with punctures with respect to the operation of attaching curves at punctures. This latter theory is also known as *quantum gravity.* A *Topological Conformal Field Theory (TCFT)* is an algebra over the chain PROP $C_*(\mathcal{P})$ for a suitable version of chains, e.g., singular. This theory is also referred to as a *string background.*

The definitions of a CFT and a TCFT are basically rewordings of those introduced by Segal [Seg04]. The definition of a CohFT is essentially a rewording of that of Kontsevich and Manin [KM94, Man99], which was perhaps motivated by Witten's paper [Wit91].

Note that a TFT may be regarded as an algebra over the PROP $H_0(\mathcal{P})$. Often, one gets a TFT by integrating over the moduli space, in which case one obtains a TFT from the top homology groups of $\mathcal{P}$ or $\overline{\mathcal{M}}$.

3.1.3 Examples

Gromov-Witten Theory

Gromov-Witten theory is basically what physicists know as a *sigma-model.* Part of the structure, essentially the genus zero part, is also known as *quantum cohomology.* Gromov-Witten theory starts with a complex projective manifold M and nonnegative integers g, m, and n and uses a diagram

$$\overline{\mathcal{M}}_{g,m,n} \xleftarrow{\ f\ } \overline{\mathcal{M}}_{g,m,n}(M;\beta) \xrightarrow{\ \mathrm{ev}\ } M^{m+n},$$

in which $\overline{\mathcal{M}}_{g,m,n}$ is the moduli space of stable compact complex curves of genus g with $m+n$ labeled punctures, the first m of them thought of as the inputs, the last n as the outputs. Here $\overline{\mathcal{M}}_{g,m,n}(M;\beta)$ is the moduli space of stable pairs (a compact complex curve C from $\overline{\mathcal{M}}_{g,m,n}$, a holomorphic map $C \to M$ taking the fundamental class of C to a given homology class $\beta \in H_2(M;\mathbb{Z})$). In the spirit of Chapter 2, the above diagram defines a correspondence, which allows one to define a pairing

$$\Omega^*(M)^{\otimes m+n} \otimes \Omega^*(\overline{\mathcal{M}}_{g,m,n}) \to \mathbb{C}[-d_{\text{virt}}],$$

given by

$$\int_{[\overline{\mathcal{M}}_{g,m,n}(M;\beta)]_{\text{virt}}} ev^*(\omega_1 \otimes \cdots \otimes \omega_{m+n}) \wedge f^*(\phi).$$

Here Ω^* refers to the complex of differential forms, [ ] denotes the degree shift, and $[\overline{\mathcal{M}}_{g,m,n}(M;\beta)]_{\text{virt}}$ is the so-called *virtual fundamental class*, a certain homology class of dimension d_{virt} of $\overline{\mathcal{M}}_{g,m,n}(M;\beta)$. This is morally the fundamental class of the ideal moduli space associated to the corresponding deformation problem. Passing to cohomology and using Poincaré duality on M, one gets a CohFT with a degree shift, see [Beh97, LT98, Man99].

Gromov-Witten Theory: a richer version

The following is just a variation on the theme of the previous example, closer to string topology and what physicists originally thought of as sigma-model. Instead of the holomorphic mapping moduli space $\overline{\mathcal{M}}_{g,m,n}(M;\beta)$, use an infinite dimensional version $\mathcal{P}_{g,m,n}(M;\beta)$ of it, the moduli space of holomorphic maps from nonsingular complex compact curves with holomorphic holes. Here the holes are not considered removed from the complex curve, so that the fundamental class of the curve is defined and mapped to a homology class β of M. As in the previous example, we get a correspondence

$$\mathcal{P}_{g,m,n} \leftarrow \mathcal{P}_{g,m,n}(M;\beta) \to (LM)^{m+n},$$

which should create something resembling a TCFT via passing to semi-infinite differential forms on the free loop space $LM = \text{Map}(S^1, M)$:

$$(\Omega^{\infty/2+*}(LM))^{\otimes m+n} \otimes \Omega^*(\mathcal{P}_{g,m,n}) \to \mathbb{C}[-d]$$

for some degree d. Much of this has not yet been made rigorous. In particular the theory of semi-infinite differential forms on the loop space has never yet been made precise. However this perspective supplies the motivation for much work in Gromov-Witten theory. For instance, Lalonde [Lal04] has described field theoretic properties along these lines with Floer homology replacing semi-infinite forms.

String topology

String topology generalizes to a TCFT in the following sense. Instead of being based on a moduli space of complex curves, it is based on a combinatorial counterpart of it, the (partial) PROP of reduced metric chord diagrams $\mathcal{RCF}_{p,q}(g)$, see next Section and also [CG04, Cha03]. Here *partial* refers to the fact that the PROP composition is only partially defined, namely, when the circumferences of the outputs and the inverse circumferences of the inputs, which one thinks of as colors, match. The correspondence realizing the string-topology TCFT at the level of correspondences is given by a diagram

$$\mathcal{RCF}_{m,n}(g) \times (LM)^m \hookleftarrow \mathcal{RCF}_{m,n}(g)M \to (LM)^n,$$

where the space in the middle is the space of continuous (or continuous, piecewise smooth, depending on the version of a free loop space LM considered) maps from reduced metric chord diagrams of genus g with m inputs and n outputs to the target manifold M. Passing to homology and using the Thom collapse map, see Chapter 1, we get a string-topology version of a CohFT given by an algebra over the homology PROP $H_*(\mathcal{RCF}_{m,n}(g))$ defined by a diagram

$$H_*(\mathcal{RCF}_{m,n}(g)) \otimes H_*(LM)^{\otimes m} \longrightarrow H_{*+\chi d}(LM)^{\otimes n},$$

where χ is the Euler characteristic of the chord diagrams of that type and $d = \dim M$, see also Equation (3.5).

3.1.4 Motivic TCFTs

The above examples suggest that one needs a more general definition of a TCFT than the one given before: the main issue being the need to use correspondences at the space level or degree shifts at the (co)homology level. Namely a *motivic TCFT* is a tensor functor from a category of holomorphic two-dimensional cobordisms (or a version thereof, such as the reduced metric chord diagrams) enriched over a motivic category of manifolds with values in a motivic category of manifolds. The tensor structure on the category of cobordisms is given by the disjoint union and on the category of manifolds by the Cartesian product. A motivic category of manifolds may be an appropriate category of correspondences, as treated earlier in Section 2.3. Motivic ideas in TCFT, long advocated by Morava, come from Kontsevich and Manin [KM94], who used a motivic axiom to describe a Cohomological Field Theory in the context of Gromov-Witten theory. Defining and studying motivic TCFTs is one of the more interesting, open projects in the study of topological field theories.

3.2 Generalized string topology operations

Recall from Chapter 1 that the loop homology product is defined by considering the diagram

$$LM \xleftarrow{\gamma} \mathrm{Map}(8, M) \xrightarrow{e} LM \times LM,$$

where $e : \mathrm{Map}(8, M) \hookrightarrow LM \times LM$ is a codimension d embedding. We used the Thom collapse map to define an umkehr map in a multiplicative generalized homology theory that supports an orientation of M,

$$e_! : h_*(LM \times LM) \to h_{*-d}(\mathrm{Map}(8, M)).$$

The loop product was defined to be the composition,

$$\mu : h_*(LM) \otimes h_*(LM) \to h_*(LM \times LM) \xrightarrow{e_!} h_{*-d}(\mathrm{Map}(8, M)) \xrightarrow{\gamma_*} h_{*-d}(LM).$$

One can think of this structure in the following way. Consider the "pair of pants" surface P, viewed as a cobordism from two circles to one circle (see Figure 3.6). We think of the two circles bounding the "legs" of the pants as incoming, and the circle bounding the "waist" as outgoing.

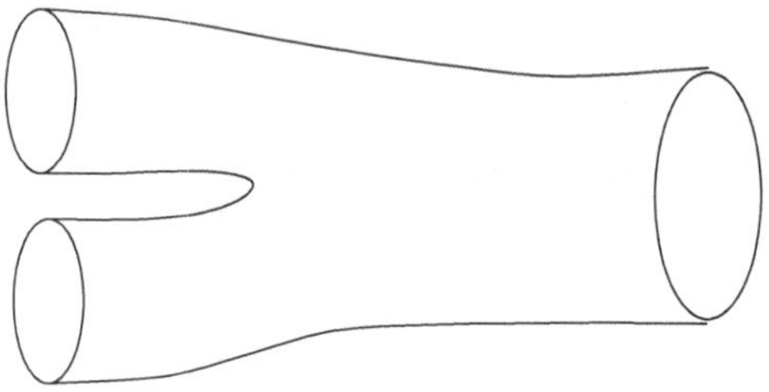

Figure 3.6: The "pair of pants" surface P

Consider the smooth mapping space, $\mathrm{Map}(P, M)$. Then there are restriction maps to the incoming and outgoing boundary circles,

$$\rho_{in} : \mathrm{Map}(P, M) \to LM \times LM, \quad \rho_{out} : \mathrm{Map}(P, M) \to LM.$$

Notice that the figure 8 is homotopy equivalent to the surface P, with respect to which the restriction map $\rho_{in} : \mathrm{Map}(P, M) \to LM \times LM$ is homotopic to the embedding $e : \mathrm{Map}(8, M) \to LM \times LM$. Also, restriction to the outgoing boundary, $\rho_{out} : \mathrm{Map}(P, M) \to LM$ is homotopic to $\gamma : \mathrm{Map}(8, M) \to LM$. So the Chas-Sullivan product can be thought of as a composition,

$$\mu_* : H_p(LM) \otimes H_q(LM) \xrightarrow{(\rho_{in})_!} H_{p+q-d}(\mathrm{Map}(P, M)) \xrightarrow{(\rho_{out})_*} H_{p+q-d}(LM).$$

The role of the figure 8 can therefore be viewed as just a technical one, that allows us to define the umkehr map $e_! = (\rho_{in})_!$.

More generally, consider a surface F, viewed as a cobordism from p-circles to q-circles. See Figure 3.7 below. Again, we think of the p boundary circles as incoming, and the remaining q boundary circles as outgoing.

We can consider the mapping space, $\mathrm{Map}(F, M)$, and the resulting restriction maps,

$$(LM)^q \xleftarrow{\rho_{out}} \mathrm{Map}(F, M) \xrightarrow{\rho_{in}} (LM)^p. \tag{3.1}$$

In [CG04] Cohen and Godin showed how to construct an umkehr map

$$(\rho_{in})_! : h_*((LM)^p) \to h_{*+\chi(F)\cdot d}(\mathrm{Map}(F, M))$$

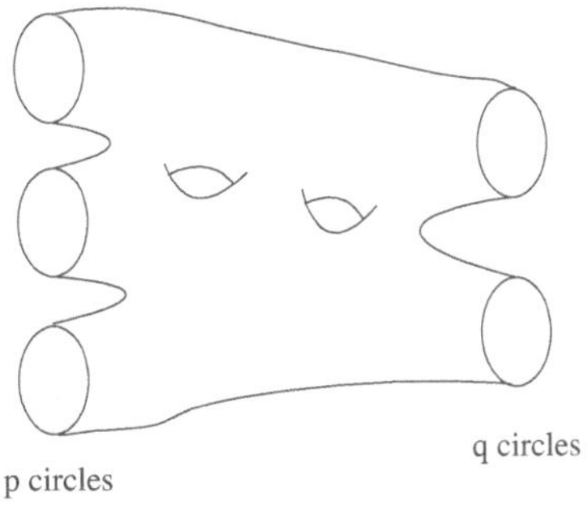

Figure 3.7: The surface F

where $\chi(F)$ is the Euler characteristic of the surface F and as above, $d = dim(M)$. This then allows the definition of a string topology operation

$$\mu_F : h_*((LM)^p) \xrightarrow{(\rho_{in})_!} h_{*+\chi(F)\cdot d}(\mathrm{Map}(F, M)) \xrightarrow{(\rho_{in})_*} h_{*+\chi(F)\cdot d}((LM)^q). \quad (3.2)$$

To construct this umkehr map, Cohen and Godin used the Chas-Sullivan idea of representing the pair of pants surface P by a figure 8, and realized the surface F by a "fat graph" (or ribbon graph). Fat graphs have been used to represent surfaces for many years, and to great success. See for example the following important works: [Har85], [Str84], [Pen87], [Kon92].

We recall the definition.

Definition 3.2.1. A fat graph is a finite graph with the following properties:

1. Each vertex is at least trivalent
2. Each vertex comes equipped with a cyclic order of the half edges emanating from it.

We observe that the cyclic order of the half edges is quite important in this structure. It allows for the graph to be "thickened" to a surface with boundary. This thickening can be thought of as assigning a "width" to the ink used in drawing a fat graph. Thus one is actually drawing a two dimensional space, and it is not hard to see that it is homeomorphic to a smooth surface. Consider the following two examples (Figure 3.8) of fat graphs which consist of the same underlying graph, but have different cyclic orderings at the top vertex.

The orderings of the edges are induced by the counterclockwise orientation of the plane. Notice that Γ_1 thickens to a surface of genus zero with four boundary components, whereas Γ_2 thickens to a surface of genus 1 with two boundary components. Of course these surfaces are homotopy equivalent, since they are each homotopy equivalent to the same underlying graph. But their diffeomorphism types are different, and that is encoded by the cyclic ordering of the vertices.

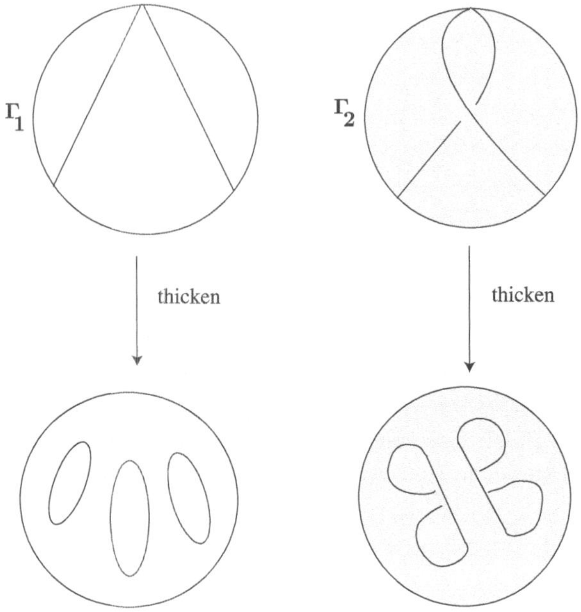

Figure 3.8: Thickenings of two fat graphs

These examples make it clear that we need to study the combinatorics of fat graphs more carefully. For this purpose, for a fat graph Γ, let $E(\Gamma)$ be the set of edges, and let $\tilde{E}(\Gamma)$ be the set of oriented edges. $\tilde{E}(\Gamma)$ is a 2-fold cover of $E(\Gamma)$. It has an involution $E \mapsto \bar{E}$, $E \in \tilde{E}(\Gamma)$, which represents changing the orientation. The cyclic orderings at the vertices determine a partition of $\tilde{E}(\Gamma)$ in the following way. Consider the example illustrated in Figure 3.9.

As above, the cyclic orderings at the vertices are determined by the counterclockwise orientation of the plane. To obtain the partition, notice that an oriented edge has well defined source and target vertices. Start with an oriented edge, and follow it to its target vertex. The next edge in the partition is the next oriented edge in the cyclic ordering at that vertex. Continue in this way until one is back at the original oriented edge. This will be the first cycle in the partition. Then continue with this process until one exhausts all the oriented edges. The resulting cycles in the partition will be called *"boundary cycles"* as they reflect the boundary circles of the thickened surface. In the case of Γ_2 illustrated in Figure 3.9, the partition into boundary cycles is given by:

$$\text{Boundary cycles of } \Gamma_2\text{:} \quad (A, B, C)\,(\bar{A}, \bar{D}, E, \bar{B}, D, \bar{C}, \bar{E}).$$

So one can compute combinatorially the number of boundary components in the thickened surface of a fat graph. Furthermore the graph and the surface have the same homotopy type, so one can compute the Euler characteristic of the

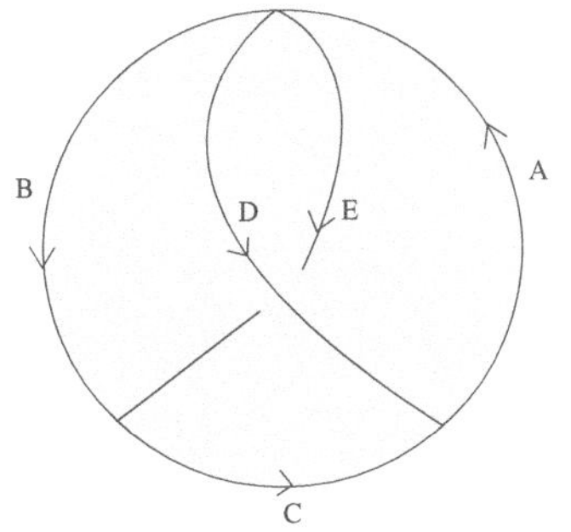

Figure 3.9: The fat graph Γ_2

surface directly from the graph. Then using the formula $\chi(F) = 2 - 2g - n$, where n is the number of boundary components, we can solve for the genus directly in terms of the graph.

A *metric fat graph* is a connected fat graph Γ endowed with a metric so that the open edges are isometrically equivalent to open intervals in the real line. Also, if $x, y \in \Gamma$, then each minimal path from x to y in Γ is isometrically equivalent to a closed interval in $\mathbb{R}$ of length $d = d(x, y)$. The set of metric fat graphs can be given a natural topology. The main theorem about these spaces is the following (see [Pen87], [Str84]).

Theorem 3.2.1. *For $g \geq 2$, the space of metric fat graphs $Fat_{g,n}$ of genus g and n boundary cycles is homotopy equivalent to the moduli space $\mathcal{M}_g^n$ of closed Riemann surfaces of genus g with n marked points.*

Notice that on a metric fat graph Γ, the boundary cycles nearly have well defined parametrizations. For example, the boundary cycle (A, B, C) of the graph Γ_2 can be represented by a map $S^1 \to \Gamma_2$ where the circle is of circumference equal to the sum of the lengths of sides A, B, and C. The ambiguity of the parametrization is the choice of where to send the basepoint $1 \in S^1$. In her thesis [God04], Godin described the notion of a "marked" fat graph, and proved the following analogue of Theorem 3.2.1.

Theorem 3.2.2. *Let $Fat^*_{g,n}$ be the space of marked metric fat graphs of genus g and n boundary components. Then there is a homotopy equivalence*

$$Fat^*_{g,n} \simeq \mathcal{M}_{g,n}$$

where $\mathcal{M}_{g,n}$ is the moduli space of Riemann surfaces of genus g having n parameterized boundary components.

In [CG04] the umkehr map $\rho_{in} : h_*((LM)^p) \to h_{*+\chi(F)\cdot d}(\mathrm{Map}(F, M))$ was constructed as follows. Let Γ be a marked fat graph representing a surface F. We will need to assume that Γ is a special kind of fat graph, which we refer to as a "Sullivan chord diagram". See Figure 3.10.

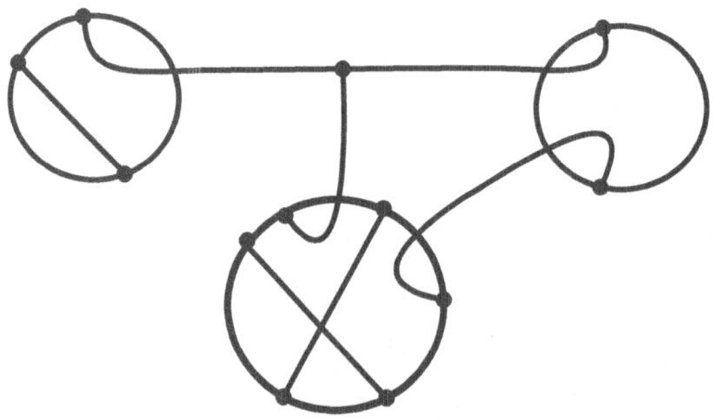

Figure 3.10: A Sullivan chord diagram

Definition 3.2.2. A "Sullivan chord diagram" of type $(g; p, q)$ is a marked fat graph representing a surface of genus g with $p+q$ boundary components, that consists of a disjoint union of p closed circles together with the disjoint union of connected trees whose endpoints lie on the circles. The cyclic orderings of the edges at the vertices must be such that each of the p disjoint circles is a boundary cycle. These p circles are referred to as the incoming boundary cycles, and the other q boundary cycles are referred to as the outgoing boundary cycles. An ordering of these boundary cycles is also part of the data.

In a chord diagram, the vertices and edges that lie on one of the p-distinct circles are called "circular vertices and edges". The noncircular edges are known as "ghost edges". As stated in the definition, the union of the ghost edges is a disjoint union of trees. Each of these trees is known as a "ghost component".

Notice that given a chord diagram c, we can construct a new fat graph $S(c)$ by collapsing all the ghost edges. That is, each ghost component is collapsed to a point. Notice that $S(c)$ is a fat graph that represents a surface of the same topological type as that represented by c (the genus and the number of boundary components are the same). However $S(c)$ is not chord diagram. We shall refer to $S(c)$ as a "reduction" of the chord diagram c. See Figure 3.11.

Figure 3.11: Reducing a chord diagram

It is not difficult to see that a Sullivan chord diagram c has the property that if an oriented edge E is contained in an incoming boundary cycle, then $\bar{E}$ is contained in an outgoing boundary cycle. In this setting the reduction $S(c)$ has the property that E is contained in an incoming boundary cycle *if and only if* $\bar{E}$ is contained in an outgoing boundary cycle. We define a reduced chord diagram to be a fat graph with this property.

Let $\mathcal{CF}_{p,q}(g)$ = space of metric Sullivan chord diagrams of topological type $(g;p,q)$. This is topologized as a subspace of $Fat_{g,p+q}$. Let $\mathcal{RCF}_{p,q}(g)$ be the corresponding space of reduced metric chord diagrams. The following facts are both due to Godin. The first is contained in [CG04], and the second is in [God04].

Proposition 3.2.3. 1. *The space $\mathcal{CF}_{p,q}(g)$ is connected.*

2. *The collapse map $\pi : \mathcal{CF}_{p,q}(g) \to \mathcal{RCF}_{p,q}(g)$ is a homotopy equivalence.*

Let $c \in \mathcal{CF}_{p,q}(g)$, and consider the mapping space, $\mathrm{Map}(S(c), M)$. This is the space of continuous maps that are smooth on each edge. Equivalently, this is the space of continuous maps $f : c \to M$, smooth on each edge, which is constant on each ghost edge. Notice that there is a homotopy equivalence $\mathrm{Map}(S(c), M) \simeq \mathrm{Map}(F_{g,p+q}, M)$, where $F_{g,p+q}$ is a surface of genus g and $p+q$ boundary components.

Markings on $S(c)$ induce parameterizations of the incoming and outgoing boundary cycles of c, so restriction to these boundary cycles induces a diagram,

$$(LM)^q \xleftarrow{\rho_{out}} \mathrm{Map}(S(c), M) \xrightarrow{\rho_{in}} (LM)^p,$$

which is homotopic to the diagram 3.1.

Recall that our goal is to define the umkehr map, $(\rho_{in})_!$. This was done in [CG04] as follows.

Let $v(c)$ be the collection of circular vertices of c. Let $\sigma(c)$ be the collection of vertices of $S(c)$. The projection $\pi : c \to S(c)$ determines a surjective set map, $\pi_* : v(c) \to \sigma(c)$. This in turn induces a diagonal map $\Delta_c : M^{\sigma(c)} \to M^{v(c)}$. Let $c_1, \cdots, c_p$ be the incoming circles in c. The markings define parameterizations and therefore identify $(LM)^p$ with $\mathrm{Map}(\coprod_{i=1}^p c_i, M)$. We then have a pullback diagram

$$\begin{array}{ccc} \mathrm{Map}_*(c, M) & \xrightarrow[\hookrightarrow]{\rho_{in}} & (LM)^p \\ {\scriptstyle e_c}\downarrow & & \downarrow{\scriptstyle e_c} \\ M^{\sigma(c)} & \xrightarrow[\Delta_c]{\hookrightarrow} & M^{v(c)} \end{array} \tag{3.3}$$

Here e_c refers to the map that evaluates at the relevant vertices.

The codimension of $\Delta_c : M^{\sigma(c)} \hookrightarrow M^{v(c)}$ is $(v(c) - \sigma(c)) \cdot d$. But a straightforward exercise verifies that $(v(c) - \sigma(c))$ is equal to minus the Euler characteristic, $v(c) - \sigma(c) = -\chi(c) = -\chi(F_{g,p+q})$. So the codimension of Δ_c is $-\chi(c) \cdot d$.

This pullback diagram allows us to construct the Thom collapse map,

$$\tau_c : (LM)^p \longrightarrow \mathrm{Map}_*(S(c), M)^{\eta_c},$$

where η_c is the normal bundle. So in homology we get an umkehr map,

$$\begin{aligned}(\rho_{in})_! : H_*((LM)^p) &\xrightarrow{(\tau_c)_*} H_*(\mathrm{Map}_*(S(c), M)^{\eta_c}) \\ &\xrightarrow[\cong]{\cap u} H_{*+\chi(c)d}(\mathrm{Map}_*(S(c), M)).\end{aligned}$$

This in turn allows us to define the string topology operation,

$$\begin{aligned}\mu_c = (\rho_{out})_* \circ (\rho_{in})_! : H_*((LM)^p) &\to H_{*+\chi(c)d}(\mathrm{Map}_*(S(c), M)) \\ &\to H_{*+\chi(c)d}((LM)^q). \end{aligned} \tag{3.4}$$

The following was proved in [CG04].

Theorem 3.2.4. *The homology operation μ_c only depends on the topology of the surface $F_{g,p+q}$. These operations can be defined for any generalized homology h_* that supports an orientation of M. They respect gluing and define a "positive boundary" topological field theory. That is, there is an operation for every surface with p incoming and q outgoing boundary components so long as $q > 0$. Equivalently, this is a Frobenius algebra whose coalgebra structure does not have a co-unit.*

The idea behind the proof of this theorem was to show that one can construct an umkehr map if one allows the chord diagram c to vary in a continuous family. The connectedness of the space of chord diagrams, $\mathcal{CF}_{p,q}(g)$ then establishes that the operation μ_c is independent of $c \in \mathcal{CF}_{p,q}(g)$. The fact that these operations respect gluing, and thereby define a field theory, uses the naturality of the Thom collapse maps. We refer the reader to [CG04] for details.

We remark that in [CG04] it was observed that by allowing c to vary over $\mathcal{CF}_{p,q}(g)$, one can actually construct operations,

$$\mu : H_*(\mathcal{CF}_{p,q}(g)) \otimes H_*(LM)^{\otimes p} \longrightarrow H_*(LM)^{\otimes q}. \tag{3.5}$$

This was verified using Jakob's bordism approach to homology theory by Chataur in [Cha03], using the language of partial PROPs. Most of the basic results about string topology described in these notes were also verified using this theory. We refer the reader to [Cha03] for details regarding this very appealing, geometric approach to string topology.

3.3 Open-closed string topology

In this section we describe work of Sullivan [Sul04] on open-closed string topology. We also discuss generalizations and expansions of this theory due to Ramirez [Ram05]. Similar constructions and results were also obtained by Harrelson in [Har04]. In this setting our background manifold comes equipped with a collection of submanifolds,

$$\mathcal{B} = \{D_i \subset M\}.$$

Such a collection is referred to as a set of "D-branes", which in string theory supplies boundary conditions for open strings. In string topology, this is reflected by considering the path spaces

$$\mathcal{P}_M(D_i, D_j) = \{\gamma : [0,1] \to M, : \gamma(0) \in D_i, \quad \gamma(1) \in D_j\}.$$

Following Segal's viewpoint, [Seg01], in a theory with D-branes, one associates to a connected, oriented, compact one-manifold S whose boundary components are labeled by D-branes, a vector space V_S. In the case of string topology, if S is topologically a circle, S^1, the vector space $V_{S^1} = h_*(LM)$. If S is an interval with boundary points labeled by D_i and D_j, then $V_S = h_*(\mathcal{P}_M(D_i, D_j))$. As is usual in field theories, to a disjoint union of such compact one-manifolds, one associates the tensor product of the vector spaces assigned to each connected component.

Now to an appropriate cobordism, one needs to associate an operator between the vector spaces associated to the incoming and outgoing parts of the boundary. In the presence of D-branes these cobordisms are cobordisms of manifolds with boundary. More precisely, in a theory with D-branes, the boundary of a cobordism F is partitioned into three parts:

1. incoming circles and intervals, written $\partial_{in}(F)$,
2. outgoing circles and intervals, written $\partial_{out}(F)$,
3. the "free part" of the boundary, written $\partial_f(F)$, each component of which is labeled by a D-brane. Furthermore $\partial_f(F)$ is a cobordism from the boundary of the incoming one-manifold to the boundary of the outgoing one-manifold. This cobordism respects the labeling.

We will call such a cobordism an "open-closed cobordism" (see Figure 3.12).

We remark that the topology of the category of open and closed strings has been evaluated by Baas, Cohen, and Ramirez in [BCR04]. This is actually a symmetric monoidal 2-category, where the objects are compact one manifolds S, whose boundary components are labeled by elements in a set of D-branes, $\mathcal{B}$. The morphisms are open-closed cobordisms, and the 2-morphisms are diffeomorphisms of these cobordisms. Let $\mathcal{S}^{oc}_{\mathcal{B}}$ denote this 2-category, and $|\mathcal{S}^{oc}_{\mathcal{B}}|$ its geometric realization. They used Tillmann's work on the category of closed strings [Til97], as well as the striking theorem of Madsen and Weiss [MW02] proving Mumford's conjecture about the stable cohomology of mapping class groups, to prove the following.

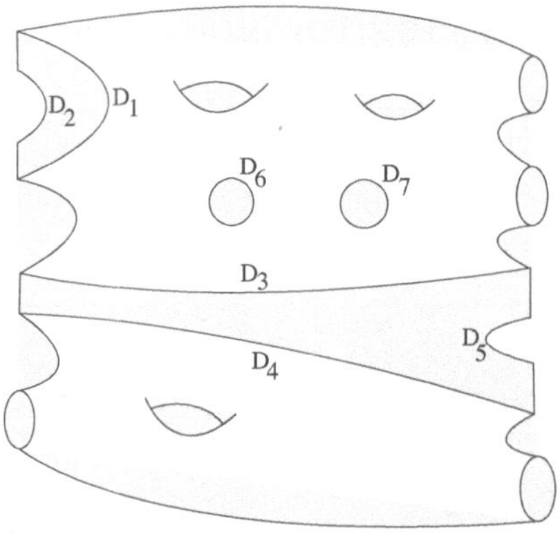

Figure 3.12: open-closed cobordism

Theorem 3.3.1. *There is a homotopy equivalence of infinite loop spaces,*

$$\Omega|\mathcal{S}_{\mathcal{B}}^{oc}| \simeq \Omega^\infty\left((\mathbb{CP}^\infty)^{-L}\right) \times \prod_{D\in\mathcal{B}} Q(\mathbb{CP}_+^\infty)$$

where $Q(Y) = \varinjlim_k \Omega^k\Sigma^k(Y)$ *and, as usual,* X_+ *denotes* X *with a disjoint basepoint. Here* $\Omega^\infty\left((\mathbb{CP}^\infty)^{-L}\right)$ *is the zero space of the Thom spectrum of the virtual bundle* $-L$*, where* $L \to \mathbb{CP}^\infty$ *is the canonical line bundle.*

In a theory with D-branes, associated to an open-closed cobordism F there is an operator,

$$\Phi_F : V_{\partial_{in}(F)} \longrightarrow V_{\partial_{out}(F)}.$$

Of course such a theory must respect gluing of open-closed cobordisms.

Such a theory with D-branes has been put into the categorical language of PROPs by Ramirez [Ram05] extending notions of Segal and Moore [Seg01]. He called such a field theory a $\mathcal{B}$-topological quantum field theory.

In the setting of string topology, operators Φ_F were defined by Sullivan [Sul04] using transversal intersections of chains. They were defined via Thom-collapse maps by Ramirez in [Ram05]. We will illustrate his definitions by the following examples.

Consider the genus zero open-closed cobordism, C_1, with free boundary components labeled by D-branes, D_1, D_2, and D_3 as indicated in Figure 3.13.

This cobordism defines an operation

$$\Phi_{C_1} : H_*(\mathcal{P}_M(D_1, D_2)) \otimes H_*(\mathcal{P}_M(D_2, D_3)) \longrightarrow H_*(\mathcal{P}_M(D_1, D_3))$$

in the following way. Let $\mathrm{Map}_{\mathcal{B}}(C_1, M)$ denote the space of smooth maps $\gamma : C_1 \to M$, where the restriction of γ to a boundary interval labeled by the D-brane

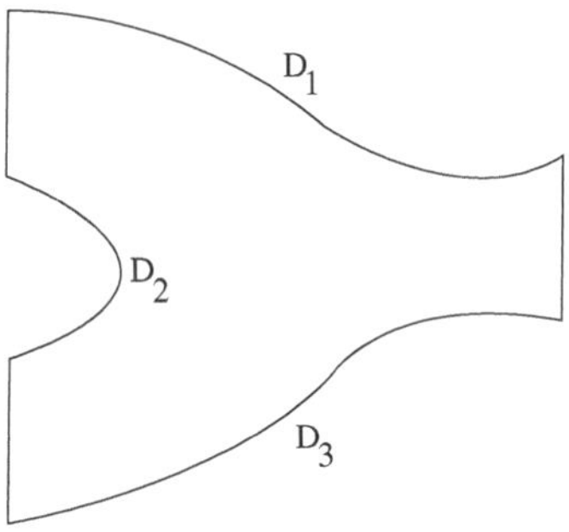

Figure 3.13: The cobordism C_1

$D_i \in \mathcal{B}$, takes values in $D_i \subset M$. Notice we have a diagram of restriction maps,

$$\mathcal{P}_M(D_1, D_3) \xleftarrow{\rho_{out}} \mathrm{Map}_{\mathcal{B}}(C_1, M) \xrightarrow{\rho_{in}} \mathcal{P}_M(D_1, D_2) \times \mathcal{P}_M(D_2, D_3) \tag{3.6}$$

As in the case of closed string operations, the main idea is to construct an umkehr map

$$(\rho_{in})_! : H_*(\mathcal{P}_M(D_1, D_2)) \otimes H_*(\mathcal{P}_M(D_2, D_3)) \to H_*(\mathrm{Map}_{\mathcal{B}}(C_1, M))$$

and then define the operation Φ_{C_1} to be the composition $\Phi_{C_1} = (\rho_{out})_* \circ (\rho_{in})_!$:

$$H_*(\mathcal{P}_M(D_1, D_2)) \otimes H_*(\mathcal{P}_M(D_2, D_3)) \to H_*(\mathrm{Map}_{\mathcal{B}}(C_1, M)) \to H_*(\mathcal{P}_M(D_1, D_3)).$$

The umkehr map $(\rho_{in})_!$ was defined in this case by replacing the mapping space $\mathrm{Map}_{\mathcal{B}}(C_1, M)$ by the path space

$$\mathcal{P}_M(D_1, D_2, D_3) = \{\alpha : [0,1] \to M \; : \; \alpha(0) \in D_1, \, \alpha(\frac{1}{2}) \in D_2, \, \alpha(1) \in D_3\}.$$

We notice that there is a restriction map $r : \mathrm{Map}_{\mathcal{B}}(C_1, M) \to \mathcal{P}_M(D_1, D_2, D_3)$ which is a homotopy equivalence. Furthermore we observe that there is a pullback diagram of fibrations,

$$\begin{array}{ccc} \mathcal{P}_M(D_1, D_2, D_3) & \longrightarrow & \mathcal{P}_M(D_1, D_2) \times \mathcal{P}_M(D_2, D_3) \\ {\scriptstyle ev_{\frac{1}{2}}} \downarrow & & \downarrow {\scriptstyle ev_1 \times ev_0} \\ D_2 & \xrightarrow[\Delta]{} & D_2 \times D_2, \end{array}$$

where the vertical fibrations are evaluation maps at the times given by the subscripts. As argued previously, the existence of this pullback square allows for the definition of a Thom collapse map,

$$\tau : \mathcal{P}_M(D_1, D_2) \times \mathcal{P}_M(D_2, D_3) \to (\mathcal{P}_M(D_1, D_2, D_3))^{\nu_\Delta} = (\mathcal{P}_M(D_1, D_2, D_3))^{TD_2},$$

which, in homology defines the umkehr map we are looking for.

We consider one more example. Consider the open-closed cobordism C_2 between an interval, whose boundary is labeled by a D-brane D, and a circle. This cobordism is pictured in Figure 3.14.

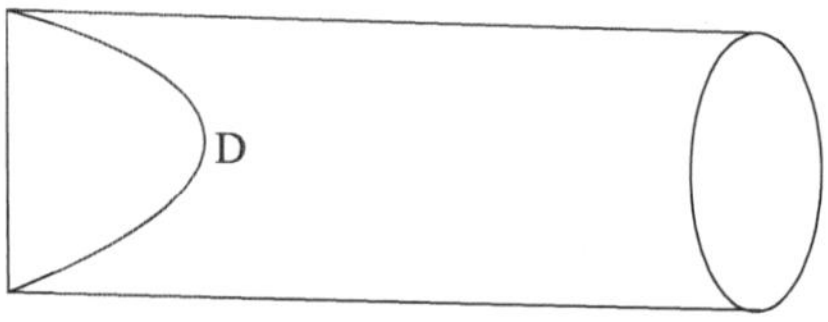

Figure 3.14: The cobordism C_2

As in the previous example, we consider the mapping space, $\mathrm{Map}_{\mathcal{B}}(C_2, M)$ consisting of maps from the surface to M, which map the free part of the boundary to $D \subset M$. Then there is a diagram of restriction maps,

$$LM \xleftarrow{\rho_{out}} \mathrm{Map}_{\mathcal{B}}(C_2, M) \xrightarrow{\rho_{in}} \mathcal{P}_M(D, D). \tag{3.7}$$

As before, the operation Φ_{C_2} will be defined to be the composition,

$$\Phi_{C_2} = (\rho_{out})_* \circ (\rho_{in})_! : H_*(\mathcal{P}_M(D, D)) \to H_*(\mathrm{Map}_{\mathcal{B}}(C_2, M)) \to H_*(LM).$$

The umkehr map $(\rho_{in})_!$ is defined as above, except that we now replace the mapping space $\mathrm{Map}_{\mathcal{B}}(C_2, M)$ by the path space

$$L_D(M) = \{\alpha \in LM \, : \, \alpha(0) \in D\}.$$

Again, there is a restriction map $r : \mathrm{Map}_{\mathcal{B}}(C_2, M) \to L_D(M)$ that is a homotopy equivalence. Furthermore, there is a pullback square,

$$\begin{array}{ccc} L_D(M) & \longrightarrow & \mathcal{P}_M(D, D) \\ {\scriptstyle ev_0}\downarrow & & \downarrow{\scriptstyle ev_0 \times ev_1} \\ D & \xrightarrow[\Delta]{} & D \times D \end{array}$$

from which we construct a Thom collapse map, $\tau : \mathcal{P}_M(D, D) \to (L_D(M))^{TD}$ and the induced umkehr map, $(\rho_{in})_! : H_*(\mathcal{P}_M(D, D)) \to H_{*-dim(D)}(L_D(M))$.

In order to construct operations Φ_Σ for an arbitrary open-closed cobordism Σ as in Figure 3.12, Ramirez replaced mapping spaces $\mathrm{Map}_{\mathcal{B}}(\Sigma, M)$ by homotopy equivalent spaces, $\mathrm{Map}_{\mathcal{B}}(\Gamma, M)$, where Γ is an appropriate fat graph with boundary labels. Ramirez defined the appropriate concept of these graphs, which he called "$\mathcal{B}$-fat graphs", studied the moduli space of such graphs, used them to construct operations Φ_Σ, and proved the following theorem [Ram05].

Theorem 3.3.2. *Given a set of D-branes $\mathcal{B}$ in a manifold M and a generalized homology theory h_* that supports orientations of M and all the submanifolds of $\mathcal{B}$, then the open-closed string topology operations define a positive boundary $\mathcal{B}$-topological quantum field theory.*

Chapter 4

A Morse theoretic viewpoint

In this chapter we describe a Morse theoretic approach to string topology and its relationship to the Floer theory of the cotangent bundle. We will survey work contained in [CN05], [Coh04a], [Coh05], [SW03a], [AS04], [Sch05].

4.1 Cylindrical gradient graph flows and string topology

As already pointed out, string topology has many of the same formal properties and structure as more geometric theories, such as Floer theory, and Gromov-Witten theory. In these theories, invariants are obtained by counting, in an appropriate sense, maps of surfaces to the manifold that satisfy the Cauchy- Riemann equations, or certain perturbations of these equations. The point of this section is to outline work in [Coh05] that shows that the string topology invariants also can be computed by counting maps of certain topological surfaces to a manifold M, that satisfy certain differential equations. The differential equations in this case are gradient flow equations of Morse functions on the loop space. In the next section we will use this point of view to describe recent work exploring relationships between string topology and Gromov-Witten theory of the cotangent bundle.

The surfaces in this study come as thickenings of fat graphs, or more specifically, marked Sullivan chord diagrams as described in chapter 3. Recall from section 2 of that chapter that a reduced Sullivan chord diagram Γ of type $(g; p, q)$ is a fat graph whose $p + q$ boundary cycles are subdivided into p incoming, and q outgoing cycles, and because of their markings, come equipped with parameterizations, which we designate by α^- for the incoming, and α^+ for the outgoing boundaries:

$$\alpha^- : \coprod_p S^1 \longrightarrow \Gamma, \quad \alpha^+ : \coprod_q S^1 \longrightarrow \Gamma.$$

If we endow Γ with a metric, we can take the circles to have circumference equal to the sum of the lengths of the edges making up the boundary cycle it parameterizes, each component of α^+ and α^- is a local isometry.

Define the surface Σ_Γ to be the mapping cylinder of these parameterizations,

$$\Sigma_\Gamma = \left(\coprod_p S^1 \times (-\infty, 0]\right) \sqcup \left(\coprod_q S^1 \times [0, +\infty)\right) \sqcup \Gamma / \sim \tag{4.1}$$

where $(t, 0) \in S^1 \times (-\infty, 0] \sim \alpha^-(t) \in \Gamma$, and $(t, 0) \in S^1 \times [0, +\infty) \sim \alpha^+(t) \in \Gamma$.

Notice that the figure 8 is a fat graph representing a surface of genus $g = 0$ and 3 boundary components. This graph has two edges, say A and B, and has boundary cycles $(A), (B), (\bar{A}, \bar{B})$. If we let (A) and (B) be the incoming cycles and $(\bar{A}, \bar{B})$ the outgoing cycle, then the figure 8 graph becomes a chord diagram. Figure 4.1 is a picture of the surface Σ_Γ, for Γ equal to the figure 8.

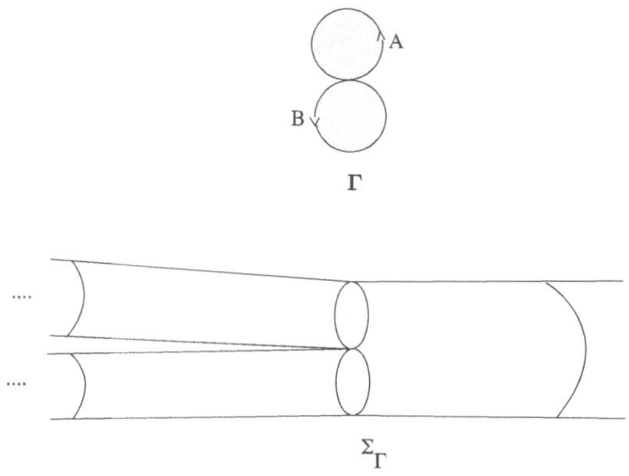

Figure 4.1: Σ_Γ

We want to study maps $\gamma : \Sigma_\Gamma \to M$. Notice that since Σ_G is made up of $p+q$ half cylinders, this is equivalent to considering p-maps, $\gamma_i : (-\infty, 0] \times S^1 \to M$, and q-maps $\gamma_j : [0, +\infty) \times S^1 \to M$ that satisfy an intersection condition at time $t = 0$ determined by the combinatorics of the fat graph Γ. Equivalently, these are maps from half lines to the loop space,

$$\gamma_i : (-\infty, 0] \to LM,\ i = 1, \cdots, p, \quad \text{and} \quad \gamma_j : [0, +\infty) \to LM,\ j = 1, \cdots, q$$

that satisfy the appropriate intersection properties at $t = 0$. We will study those maps $\gamma : \Sigma_\Gamma \to M$ so that the curves γ_i and γ_j satisfy the gradient flow lines of certain Morse functions on LM..

The constructions we are about to describe are motivated by the use of moduli spaces of "gradient graph flows" that have been used to define classical cohomology operations on compact manifolds in [BC94], [Fuk93], and [CN05]. We also refer the reader to [Coh04a] for a general description of this theory, as well as its adaptations to string topology.

Assume M is endowed with a Riemannian metric, and LM has the induced L^2-metric. Let $f : LM \to \mathbb{R}$ be a Morse function on the loop space, which is bounded below, and that satisfies the Palais-Smale condition. Recall that this condition states that if $\{a_i\}$ is a sequence of points on which the gradient tends to zero, $\lim_{i\to\infty} \nabla f(a_i) = 0$, then there is a subsequence that converges to a critical point.

Recall that given such a function f, a gradient flow line of f is a curve $\gamma : \mathbb{R} \to LM$ that satisfies the gradient flow equation:

$$\frac{d\gamma}{dt} + \nabla f(\gamma(t)) = 0. \tag{4.2}$$

If $a \in LM$ is a critical point of f, let $W^u(a)$ be the unstable manifold, and $W^s(a)$ be the stable manifold. Recall that $x \in W^u(a)$ if and only if there is a gradient flow $\gamma : \mathbb{R} \to LM$ satisfying the initial condition, $\gamma(0) = x$, and the asymptotic condition, $\lim_{t\to-\infty} \gamma(t) = a$. The stable manifold $W^s(a)$ is defined similarly except the boundary condition is that $\lim_{t\to+\infty} \gamma(t) = a$.

In the case we are considering, ($f : LM \to \mathbb{R}$ Morse, bounded below, and satisfies the Palais-Smale condition), then $W^u(a)$ is diffeomorphic to a disk of dimension $\lambda(a)$ called the index of a, and $W^s(a)$ is infinite dimensional, but has codimension equal to $\lambda(a)$. When, in addition, f satisfies the Morse-Smale transversality conditions, i.e the unstable and stable manifolds intersect transversally, one can consider the moduli spaces of gradient flow lines,

$$\mathcal{M}(a,b) = W^u(a) \cap W^s(b)/\mathbb{R}$$

where $\mathbb{R}$ acts by reparameterizing the flow lines. This is a manifold of dimension $\lambda(a) - \lambda(b) - 1$.

In this case there is a CW complex homotopy equivalent to LM, which is built out of one cell of dimension p for every critical point a of index $\lambda(a) = p$. The cellular chain complex is referred to as the Morse complex, $C_*^f(LM)$:

$$\to \cdots \xrightarrow{\partial_{p+1}} C_p^f(LM) \xrightarrow{\partial_p} C_{p-1}^f(LM) \xrightarrow{\partial_{p-1}} \cdots \tag{4.3}$$

Here $C_p^f(LM)$ is the free abelian group generated by the critical points of index p. As usual, if M is oriented, the boundary homomorphism can be computed by the formula

$$\partial_p(a) = \sum_{\lambda(b)=p-1} n_{a,b} \cdot b$$

where $n_{a,b} = \#\mathcal{M}(a,b)$. This number is counted with sign, which makes sense because in this case $\mathcal{M}(a,b)$ is a compact, oriented, zero dimensional manifold. We will let $C_f^*(LM)$ denote the dual cochain complex for computing $H^*(LM)$.

We now discuss a plentiful supply of Morse functions on LM. Consider a potential function on M, defined to be a smooth map

$$V : \mathbb{R}/\mathbb{Z} \times M \longrightarrow \mathbb{R}.$$

We can then define the classical energy functional

$$\begin{aligned} \mathcal{S}_V : LM &\longrightarrow \mathbb{R} \\ \gamma &\longrightarrow \int_0^1 \left(\frac{1}{2} \left| \frac{d\gamma}{dt} \right|^2 - V(t, \gamma(t)) \right) dt. \end{aligned} \tag{4.4}$$

For a generic choice of V, $\mathcal{S}_V$ is a Morse function [Web02] satisfying the Palais-Smale condition. Its critical points are those $\gamma \in LM$ satisfying the ODE

$$\nabla_t \frac{d\gamma}{dt} = -\nabla V_t(x) \tag{4.5}$$

where $\nabla V_t(x)$ is the gradient of the function $V_t(x) = V(t, x)$, and $\nabla_t \frac{d\gamma}{dt}$ is the Levi-Civita covariant derivative.

We will now describe a metric and Morse theoretic structure on a marked chord diagram Γ. This will allow us to define the differential equations that we would like the maps $\Sigma_\Gamma \to M$ to satisfy.

Definition 4.1.1. Given a marked chord diagram Γ with p-incoming and q-outgoing boundary cycles, we define an LM-Morse structure σ on Γ to be a metric on Γ together with a labeling of each boundary cycle of Γ by a distinct Morse function $f : LM \to \mathbb{R}$ which is bounded below, and that satisfies the Palais-Smale condition.

Notice that we can think of such a labeling of boundary cycles, as a labeling of the boundary *cylinders* of the surface Σ_Γ. Notice also that we can choose our Morse functions to be energy functions of the sort mentioned above, in which case the labeling can be taken to be by potential functions, $V : \mathbb{R}/\mathbb{Z} \times M \to \mathbb{R}$.

This leads to the following definition of the moduli space of cylindrical flows.

Definition 4.1.2. Let Γ be a marked chord diagram as above. Let σ be a LM-Morse structure on Γ. Suppose $\phi : \Sigma_\Gamma \to M$ is a continuous map, smooth on the cylinders. Let $\phi_i : S^1 \times (-\infty, 0] \to M$ be the restriction of ϕ to the ith incoming cylinder, $i = 1, \cdots, p$, and $\phi_j : S^1 \times [0, +\infty) \to M$ be the restriction to the jth outgoing cylinder, $j = 1, \cdots, q$. We consider the ϕ_i's and ϕ_j's as curves in the loop space, LM. Then the moduli space of cylindrical flows is defined to be

$$\mathcal{M}_\Gamma^\sigma(LM) = \{\phi : \Sigma_\Gamma \to M : \frac{d\phi_i}{dt} + \nabla f_i(\phi_i(t)) = 0 \quad \text{and} \quad \frac{d\phi_j}{dt} + \nabla f_j(\phi_j(t)) = 0 \\ \text{for} \quad i = 1, \cdots, p \quad \text{and} \quad j = 1, \cdots, q.\}$$

Let $\phi \in \mathcal{M}^{\sigma}_{\Gamma}(LM)$. For $i = 1, \cdots, p$, let $\phi_{i,-1} : S^1 \to M$ be the restriction of $\phi_i : S^1 \times (-\infty, 0] \to M$ to $S^1 \times \{-1\}$. Similarly, for $j = 1, \cdots, q$, let $\phi_{j,1} : S^1 \to M$ be the restriction of ϕ_j to $S^1 \times \{1\}$. These restrictions define the following maps.

$$(LM)^q \xleftarrow{\rho_{out}} \mathcal{M}^{\sigma}_{\Gamma}(LM) \xrightarrow{\rho_{in}} (LM)^p. \tag{4.6}$$

In [Coh05] it is shown that one can define a Thom collapse map,

$$\tau_\Gamma : (LM)^p \to (\mathcal{M}^{\sigma}_{\Gamma}(LM)^{\nu}$$

where ν is a certain vector bundle of dimension $-\chi(\Gamma) \cdot d$. This can be thought of as a normal bundle in an appropriate sense. This allows the definition of an umkehr map

$$(\rho_{in})_! : h_*((LM)^p) \to h_{*+\chi(\Gamma)\cdot d}(\mathcal{M}^{\sigma}_{\Gamma}(LM)) \tag{4.7}$$

for any homology theory h_* supporting an orientation of M. One can then define an operation

$$q^{morse}_{\Gamma} : h_*((LM)^p) \xrightarrow{(\rho_{in})_!} h_{*+\chi(\Gamma)\cdot d}(\mathcal{M}^{\sigma}_{\Gamma}(LM)) \xrightarrow{(\rho_{out})_*} h_{*+\chi(\Gamma)\cdot d}((LM)^q). \tag{4.8}$$

Consider the inclusion map, $j : \mathcal{M}^{\sigma}_{\Gamma}(LM) \hookrightarrow \mathrm{Map}(\Sigma_G, M)$. With respect to this map, diagram 4.6 is the restriction of diagram 3.1 described in the last chapter. Furthermore, the construction of the Thom collapse map τ_G and resulting umkehr map $(\rho_{in})_!$ in [Coh05] is compatible with the corresponding Thom collapse map and umkehr map from [CG04] described in the last chapter. This then yields the following theorem, proved in [Coh05].

Theorem 4.1.1. *For any marked chord diagram Γ, the Morse theoretic operation*

$$q^{morse}_{\Gamma} : h_*((LM)^p) \longrightarrow h_{*+\chi(\Gamma)\cdot d}((LM)^q)$$

given in (4.8) *is equal to the string topology operation*

$$q_\Gamma : h_*((LM)^p) \longrightarrow h_{*+\chi(\Gamma)\cdot d}((LM)^q)$$

defined in Theorem 3.2.4.

This Morse theoretic viewpoint of the string topology operations has another, more geometric interpretation due to Ramirez [Ram05]. It is a direct analogue of the perspective on the graph operations in [BC94].

As above, let Γ be a marked chord diagram, and σ be an LM-Morse structure on Γ. Let $(f_1, \cdots, f_{p+q})$ be the Morse functions on LM labeling the $p+q$ cylinders of Σ_Γ. As above, the first p of these cylinders are incoming, and the remaining q are outgoing.

Let $\vec{a} = (a_1, \cdots, a_{p+q})$ be a sequence of loops such that $a_i \in LM$ is a critical point of $f_i : LM \to \mathbb{R}$. Then define

$\mathcal{M}^{\sigma}_{\Gamma}(LM, \vec{a}) = \{\phi : \Sigma_\Gamma \to M$ that satisfy the following two conditions:

1. $\frac{d\phi_i}{dt} + \nabla f_i(\phi_i(t)) = 0$ for $i = 1, \cdots, p+q$
2. $\phi_i \in W^u(a_i)$ for $i = 1, \cdots, p$, and $\phi_j \in W^s(a_j)$ for $j = p+1, \cdots p+q.\}$

Ramirez then proved that, under sufficient transversality conditions described in [Ram05], $\mathcal{M}^{\sigma}_{\Gamma}(LM, \vec{a})$ is a smooth manifold of dimension

$$dim(\mathcal{M}^{\sigma}_{\Gamma}(LM, \vec{a})) = \sum_{i=1}^{p} Ind(a_i) - \sum_{j=p+1}^{p+q} Ind(a_j) + \chi(\Gamma) \cdot d. \tag{4.9}$$

Moreover, an orientation on M induces an orientation on $\mathcal{M}^{\sigma}_{\Gamma}(LM, \vec{a})$. Furthermore compactness issues are addressed, and it is shown that if the dimension $dim(\mathcal{M}^{\sigma}_{\Gamma}(LM, \vec{a}) = 0$ then it is compact. This leads to the following definition. For f_i one of the labeling Morse functions, let $C^{f_i}_*(LM)$ be the Morse chain complex for computing $H_*(LM)$, and let $C^*_{f_i}(LM)$ be the corresponding cochain complex. Consider the chain

$$\begin{aligned} q^{morse}_{\Gamma}(LM) &= \sum_{dim(\mathcal{M}^{\sigma}_{\Gamma}(LM, \vec{a}))=0} \#\mathcal{M}^{\sigma}_{\Gamma}(LM, \vec{a}) \cdot [\vec{a}] \\ &\in \bigotimes_{i=1}^{p} C^*_{f_i}(LM) \otimes \bigotimes_{j=p+1}^{p+q} C^{f_j}_*(LM) \end{aligned} \tag{4.10}$$

We remark that the (co)chain complexes $C^*_{f_i}(LM)$ are generated by critical points, so this large tensor product of chain complexes is generated by vectors of critical points $[\vec{a}]$. It is shown in [Ram05] that this chain is a cycle and if one uses (arbitrary) field coefficients this defines a class

$$\begin{aligned} q^{morse}_{\Gamma}(LM) &\in (H^*(LM))^{\otimes p} \otimes (H_*(LM))^{\otimes q} \\ &= \mathrm{Hom}((H_*(LM))^{\otimes p}, (H_*(LM))^{\otimes q}). \end{aligned} \tag{4.11}$$

Ramirez then proved that these operations are the same as those defined by (4.8), and hence by Theorem 4.1.1 is equal to the string topology operation. In the case when Γ is the figure 8, this operation is the same as that defined by Abbondandolo and Schwarz [AS04] in the Morse homology of the loop space.

4.2 Cylindrical holomorphic curves in T^*M

This section is somewhat speculative. It is based on conversations with Y. Eliashberg and is motivated by the work of Salamon and Weber [SW03a]. The goal of the work described in this section can be divided into two parts.

1. In the previous section, string topology operations were defined in terms of the moduli space of cylindrical graph flows, $\mathcal{M}^{\sigma}_{\Gamma}(LM)$. Here Γ is a marked

chord diagram, and σ is an LM-Morse structure on Γ (see definitions 4.1.1 and 4.1.2.) These consisted of maps $\gamma : \Sigma_\Gamma \to M$, that satisfy appropriate gradient flow equations on the cylinders, dictated by the structure σ. We would like to replace this moduli space by a space of maps to the cotangent bundle, $\phi : \Sigma_\Gamma \to T^*M$ that satisfy appropriate Cauchy-Riemann equations when restricted to the cylinders. These equations are determined by the structure σ, and an almost complex structure on T^*M.

2. We would like to understand how string topology type invariants defined using these moduli spaces of "cylindrical holomorphic curves", are related to invariants such as the Gromov-Witten invariants, which are defined using moduli spaces of holomorphic curves from a Riemann surface.

We will just give outlines of the ideas of this program in this section. This program is described in more detail in [Coh04a].

The fact that the cotangent bundle T^*M has an almost complex structure comes from the existence of its canonical symplectic structure. This structure is defined as follows:

Let $p : T^*M \to M$ be the projection map. Let $x \in M$ and $u \in T^*_xM$. Consider the composition

$$\alpha(x,u) : T_{(x,u)}(T^*(M)) \xrightarrow{Dp} T_xM \xrightarrow{u} \mathbb{R}$$

where $T_{(x,u)}(T^*(M))$ is the tangent space of $T^*(M)$ at (x,u), and Dp is the derivative of p. Notice that α is a one form, $\alpha \in \Omega^1(T^*(M))$, and we define

$$\omega = d\alpha \in \Omega^2(T^*(M)).$$

The form ω is a nondegenerate symplectic form on $T^*(M)$. Now given a Riemannian metric on M, $g : TM \xrightarrow{\cong} T^*M$, one gets a corresponding almost complex structure J_g on $T^*(M)$ defined as follows.

The Levi-Civita connection defines a splitting of the tangent bundle of the cotangent bundle $T^*(M)$,

$$T(T^*(M)) \cong p^*(TM) \oplus p^*(T^*(M)).$$

With respect to this splitting, $J_g : T(T^*(M)) \to T(T^*(M))$ is defined by the matrix,

$$J_g = \begin{pmatrix} 0 & -g^{-1} \\ g & 0 \end{pmatrix}.$$

The induced metric on $T^*(M)$ is defined by

$$G_g = \begin{pmatrix} g & 0 \\ 0 & g^{-1} \end{pmatrix}.$$

Now let Γ be a marked chord diagram, and σ an LM-Morse structure on Γ. Recall that this consists of a metric on Γ and a labeling of the boundary cycles by

Morse functions $f_i : LM \to \mathbb{R}$. We now assume that these Morse functions are of the form

$$f_i = S_{V_i} : LM \to \mathbb{R}$$

where S_{V_i} is the energy functional given in definition 4.4 using a potential function $V_i : \mathbb{R}/\mathbb{Z} \times M \to \mathbb{R}$.

In [SW03a] Salamon and Weber showed how, given such a potential function V, one can define a Hamiltonian function on the cotangent bundle, $H_V : \mathbb{R}/\mathbb{Z} \times T^*(M) \to \mathbb{R}$ by the formula

$$H_V(t, (x, u)) = \frac{1}{2}|u|^2 + V(t, x). \tag{4.12}$$

Using this Hamiltonian, Salamon and Weber studied the perturbed symplectic action functional on the loop space of the cotangent bundle,

$$\mathcal{A}_V : L(T^*M) \to \mathbb{R} \tag{4.13}$$

$$(\gamma, \eta) \to \mathcal{A}(\gamma, \eta) - \int_0^1 H(t, (\gamma(t), \eta(t)))dt. \tag{4.14}$$

Here (γ, η) represents a loop in T^*M, where $\gamma \in LM$ and $\eta(t) \in T^*_{\gamma(t)}M$. The classical symplectic action $\mathcal{A} : L(T^*M) \to \mathbb{R}$ is defined by

$$\mathcal{A}(\gamma, \eta) = \int_0^1 \langle \eta(t), \frac{d\gamma}{dt}(t) \rangle dt.$$

Following Floer's original construction, one can define a "Floer complex", $CF_*^V(T^*M)$, generated by the critical points of $\mathcal{A}_V$, and whose boundary operator is defined by counting gradient flow lines of $\mathcal{A}_V$. As shown in [SW03a], these are curves $(u, v) : \mathbb{R} \to L(T^*M)$, or equivalently,

$$(u, v) : \mathbb{R} \times S^1 \to T^*M$$

that satisfy the perturbed Cauchy Riemann equations,

$$\partial_s u - \nabla_t v - \nabla V_t(u) = 0 \quad \text{and} \quad \nabla_s v + \partial_t u - v = 0. \tag{4.15}$$

We refer to these maps as holomorphic cylinders in T^*M with respect to the almost complex structure J_g and the Hamiltonian H_V.

Salamon and Weber also observed that the critical points of $\mathcal{A}_V$ are loops (γ, η), where $\gamma \in LM$ is a critical point of the energy functional $\mathcal{S}_V : LM \to \mathbb{R}$, and η is determined by the derivative $\frac{d\gamma}{dt}$ via the metric, $\eta(v) = \langle v, \frac{d\gamma}{dt} \rangle$. Thus the critical points of $\mathcal{A}_V$ and those of $\mathcal{S}_V$ are in bijective correspondence. The following result is stated in a form proved by Salamon and Weber in [SW03a], but the conclusion of the theorem was first proved by Viterbo [Vit96].

Theorem 4.2.1. *The Floer chain complex $CF_*^V(T^*M)$ and the Morse complex $C_*^V(LM)$ are chain homotopy equivalent. There is a resulting isomorphism of the Floer homology of the cotangent bundle with the homology of the loop space,*

$$HF_*^V(T^*M) \cong H_*(LM).$$

This result was also proved using somewhat different methods by Abbondandolo and Schwarz [AS04].

The Salamon-Weber argument involved scaling the metric on M, $g \to \frac{1}{\epsilon}g$, which scales the almost complex structure $J \to J_\epsilon$, and the metric on T^*M, $G \to G_\epsilon = \begin{pmatrix} \frac{1}{\epsilon}g & 0 \\ 0 & \epsilon g^{-1} \end{pmatrix}$. Notice that in this metric, the "vertical" distance in the cotangent space is scaled by ϵ.

Salamon and Weber proved that there is an $\epsilon_0 > 0$ so that for $\epsilon < \epsilon_0$, the set of these holomorphic cylinders defined with respect to the metric G_ϵ, that connect critical points (a_1, b_1) and (a_2, b_2) where a_1 and a_2 have relative Morse index one with respect to the action functional $\mathcal{S}_V$, (or equivalently (a_1, b_1) and (a_2, b_2) have relative Conley-Zehnder index one) is in bijective correspondence with the set of gradient trajectories of the energy functional $\mathcal{S}_V : LM \to \mathbb{R}$ defined with respect to the metric $\frac{1}{\epsilon}g$ that connect a_1 to a_2. Theorem 4.2.1 is then a consequence.

Now again consider a marked chord diagram Γ with an LM Morse structure σ whose labeling Morse functions are of the form $f_i = \mathcal{S}_{V_i}$ for some potential $V_i : \mathbb{R}/\mathbb{Z} \times M \to \mathbb{R}$. Using the Salamon-Weber idea, we replace the moduli space of cylindrical graph flows, $\mathcal{M}_\Gamma^\sigma(LM)$, which consists of functions $\gamma : \Sigma_G \to M$ so that the restriction to the ith boundary cylinder is a gradient trajectory of the classical energy functional $\mathcal{S}_{V_i}$, by the moduli space of "cylindrical holomorphic curves" in the cotangent bundle $T^*(M)$, $\mathcal{M}^{hol}_{(\Gamma,\sigma,\epsilon)}(T^*M)$, which consists of continuous maps,

$$\phi : \Sigma_\Gamma \to T^*(M)$$

such that the restrictions to the cylinders,

$$\phi_i : (-\infty, 0] \times S^1_{c_i} \to T^*M \quad \text{and} \quad \phi_j : [0, +\infty) \times S^1_{c_j} \to T^*M$$

are holomorphic with respect to the almost complex structure J_ϵ and the Hamiltonians H_{V_i} and H_{V_j} respectively. Here the circles S^1_c are round with circumference c_j determined by the metric given by the structure σ.

Like in the last section we have restriction maps (compare (4.6))

$$(L(T^*M))^q \xleftarrow{\rho_{out}} \mathcal{M}^{hol}_{(\Gamma,\sigma,\epsilon)}(T^*M) \xrightarrow{\rho_{in}} (L(T^*M))^p. \tag{4.16}$$

ρ_{in} is defined by sending a cylindrical flow ϕ to $\prod_{i=1}^p \phi_{i,-1} : \{-1\} \times S^1 \to T^*M$ and ρ_{out} sends ϕ to $\prod_{j=p+1}^{p+q} \phi_{j,1} : \{1\} \times S^1 \to T^*M$.

Motivated by the umkehr maps $(\rho_{in})_! : h_*((LM)^p) \to h_{*+\chi(\Gamma)\cdot d}(\mathcal{M}_\Gamma^\sigma(LM))$ given in the last section, as well as the Salamon-Weber results, we conjecture the following analogue of the existence of the string topology operations, and their field theoretic properties.

Conjecture 4.2.2. *For every marked chord diagram* Γ, *there is an umkehr map*

$$(\rho_{in})_! : (HF_*(T^*M))^{\otimes p} \to H_{*+\chi(\Gamma)\cdot d}(\mathcal{M}^{hol}_{(\Gamma,\sigma,\epsilon)}(T^*M))$$

and a homomorphism

$$(\rho_{out})_* : H_*(\mathcal{M}^{hol}_{(\Gamma,\sigma,\epsilon)}(T^*M)) \to (HF_*(T^*M))^{\otimes q}$$

so that the operations

$$\theta_\Gamma = (\rho_{out})_* \circ (\rho_{in})_! : (HF_*(T^*M))^{\otimes p} \to (HF_*(T^*M))^{\otimes q}$$

satisfy the following properties:

1. *The maps* θ *fit together to define a positive boundary, topological field theory.*
2. *With respect to the Salamon-Weber isomorphism* $HF_*(T^*M) \cong H_*(LM)$ *(Theorem 4.2.1) the Floer theory operations* θ_Γ *equal the string topology operations* q_Γ *studied in the last two sections.*

Remark. The existence of a field theory structure on the Floer homology of a closed symplectic manifold was established by Lalonde [Lal04]. The above conjecture should be directly related to Lalonde's constructions.

Now one might also take the more geometric approach to the construction of these Floer theoretic operations, analogous to Ramirez's geometrically defined Morse theoretic constructions of string topology operations. This would involve the study of the space of cylindrical holomorphic curves in T^*M, with boundary conditions in stable and unstable manifolds of critical points, $\mathcal{M}^{hol}_{(\Gamma,\sigma,\epsilon)}(T^*M, \vec{a})$. Smoothness and compactness properties need to be established for these moduli spaces. In particular, in a generic situation their dimensions should be given by the formula

$$dim\,(\mathcal{M}^{hol}_{(\Gamma,\sigma,\epsilon)}(T^*M,\vec{a})) = \sum_{i=1}^{p} Ind(a_i) - \sum_{j=p+1}^{p+q} Ind(a_j) + \chi(\Gamma)\cdot d$$

where $Ind(a_i)$ denotes the Conley-Zehnder index.

We remark that in the case of the figure 8, this analysis has all been worked out by Abbondandolo and Schwarz [Sch05]. In this case Σ_Γ is a Riemann surface structure on the pair of pants. They proved the existence of a "pair of pants" algebra structure on $HF^V_*(LM)$ and with respect to their isomorphism, $HF^V_*(LM) \cong H_*(LM)$ it is isomorphic to the pair of pants product on the Morse homology of LM. In view of the comment following Definition 4.1.1 we have the following consequence.

Theorem 4.2.3. *With respect to the isomorphism $HF_*^V(T^*M) \cong H_*(LM)$, the pair of pants product in the Floer homology of the cotangent bundle corresponds to the Chas-Sullivan string topology product.*

Another aspect of the relationship between the symplectic structure of the cotangent bundle and the string topology of the manifold, has to do with the relationship between the moduli space of J-holomorphic curves with cylindrical boundaries, $\mathcal{M}_{g,n}(T^*M)$, and moduli space of cylindrical holomorphic curves, $\mathcal{M}^{hol}_{(\Gamma,\sigma,\epsilon)}(T^*M)$, where we now let Γ and σ vary over the appropriate space of metric graphs. We conjecture that these moduli spaces are related as a parameterized version of the relationship between the moduli space of Riemann surfaces and the space of metric fat graphs (Theorem 3.2.1).

Once established, this conjecture would give a direct relationship between Gromov-Witten invariants of the cotangent bundle, and the string topology of the underlying manifold. In this setting the Gromov-Witten invariants would be defined using moduli spaces of curves with cylindrical ends rather than marked points, so that the invariants would be defined in terms of the homology of the loop space (or, equivalently, the Floer homology of the cotangent bundle), rather than the homology of the manifold.

We believe that there is a very deep relationship between the symplectic topology of the cotangent bundle and the string topology of the underlying manifold. There is much work yet to be done in understanding the extent of this relationship.

Chapter 5

Brane topology

5.1 The higher-dimensional cacti operad

String topology may be generalized to higher-dimensional sphere spaces $SM := M^{S^n} = \mathrm{Map}(S^n, M)$ for $n \geq 1$, see [SV05]. See also the paper [Cha03] by Chataur, in which a string (i.e., "dot") product on the homology of sphere spaces M^{S^n} was defined, as well as the other string topology operations for $n = 3$. In the paper [KS03] by Kallel and Salvatore, the string product for $n = 2$ and $M = \mathbb{C}P^N$ was shown to be coming from the holomorphic mapping space $\mathrm{Hol}(\mathbb{C}P^1, \mathbb{C}P^N)$ via Segal's homological approximation theorem of the continuous maps by holomorphic ones.

The corresponding cacti operad does not admit a nice combinatorial description available for $n = 1$, so that we will consider an awfully big, but neat, infinite dimensional operad, which will do the job.

For $n \geq 1$, the *n-dimensional cacti operad* is an operad of topological spaces. It may be described as a collection of topological spaces $\mathcal{C}^n(k)$, $k \geq 1$, defined as follows. The space $\mathcal{C}^n(k)$ is the space of all continuous maps from the unit n-sphere S^n to the union of k labeled (by the numbers $1, \dots, k$) n-spheres, called the *lobes*, joined at a few points, such that every two lobes intersect at most at one point and the *dual graph* (whose vertices are the lobes and the intersection points and whose edges connect the lobes with the adjacent intersection points) of this union is a connected tree. One can think of a point of the space $\mathcal{C}^n(k)$ as a pair (c, ϕ), where c is a join of k spheres, as above, called a *cactus*, see Figure 5.1, and $\phi : S^n \to c$ a continuous map, called a *structure map* from the unit n-sphere to the cactus.

The topology on $\mathcal{C}^n(k)$ is given as follows. The topology on the set of cacti c is induced from the following inclusion into $(S^n)^{k(k-1)}$: for each ordered pair (i, j), $1 \leq i \neq j \leq k$, one takes the point on the ith lobe of a given cactus at which this lobe is attached to a lobe eventually leading to the jth lobe. The topology on the "*universal cactus*" $\mathcal{U}$ defined as the set of pairs (c, x), where c is a cactus and x is

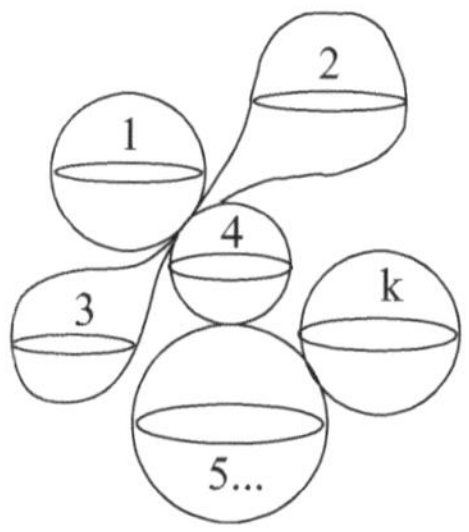

Figure 5.1: A 2d-cactus.

a point on it, is induced from the inclusion of $\mathcal{U}$ into $(S^n)^{k^2} = (S^n)^{k(k-1)} \times (S^n)^k$, the second factor governing the position of x on c from the point of view of each lobe. The universal cactus projects naturally onto the *space of cacti*, the space of tree-like joins of k labeled spheres, with the "fiber" being the corresponding cactus. Finally, a basis of the topology on the set $\mathcal{C}^n(k)$, which may be identified with the set of continuous maps from the unit sphere S^n to the fibers of the universal cactus, is formed by finite intersections of the sets of such maps $\Phi : S^n \to \mathcal{U}$ satisfying $\Phi(K) \subset U$ for given compact $K \subset S^n$ and open $U \subset \mathcal{U}$.

An operad structure on the collection $\mathcal{C}^n = \{\mathcal{C}^n(k) \mid k \geq 1\}$ is defined as follows. An action of the symmetric group is defined via changing the labels of the lobes. A unit element $\mathrm{id} \in \mathcal{C}^n(1)$ is the identity map $\mathrm{id} : S^n \to S^n$. Operad compositions $\circ_i : \mathcal{C}^n(k) \times \mathcal{C}^n(l) \to \mathcal{C}^n(k+l-1)$, $i = 1, \ldots, k$, are defined for two cacti C_1 with k lobes and C_2 with l lobes as follows. First, construct a new cactus by attaching C_1 to C_2 via a certain attaching map from the ith lobe of C_1 to the cactus C_2. This attaching map is nothing but the structure map f_2 from the unit sphere to the cactus C_2. At the level of dual graphs, we glue in the tree corresponding to C_2 in place of the vertex corresponding to the ith lobe of C_1 by connecting the incoming edges into this vertex in C_1 with certain vertices of the tree of C_2, as prescribed by the map f_2. This procedure, similar to ones appearing in [CK00, KS00], results in a tree again at the level of dual graphs and gives a cactus $C_1 \circ_i C_2 := C_2 \cup_{f_2} C_1$ with $k+l-1$ lobes. Label the lobes by inserting the labels $1, \ldots, l$ of C_2 at the ith space in the set of labels $1, \ldots, k$ of C_1 and relabeling the resulting linearly ordered set as $1, \ldots, k+l-1$. Then, define a structure map for this cactus as the composition of the structure map $f_1 : S^n \to C_1$ and the natural map $C_1 \to C_2 \cup_{f_2} C_1$.

This operad is a close relative of the *framed little* $(n+1)$*-disks operad* $f\mathcal{D}^{n+1}$, see Section 2.1.4. Little is known about a direct relation, except the following theorem and Remark 4 of Salvatore, which one might try to generalize to higher-dimensional cacti.

Theorem 5.1.1 (Sullivan-AV [SV05]). *There exists an operad morphism*

$$H_*(f\mathcal{D}^{n+1}; \mathbb{Q}) \to H_*(\mathcal{C}^n; \mathbb{Q})$$

from the rational homology framed little $(n+1)$-disks operad to the rational homology n-cacti operad.

Idea of proof. One constructs maps $f\mathcal{D}^n(k) \to \mathcal{C}^n(k)$, $k = 1, 2, 3$, from the first three components of the little disks operad to the cacti operad. To do that, first construct $\mathrm{SO}(n+1)$-equivariant maps $\mathcal{D}^{n+1}(k) \to \mathcal{C}^n(k)$, $k = 1, 2, 3$, for the (non-framed) little disks operad $\mathcal{D}^{n+1}$ and then extend them to $f\mathcal{D}^{n+1}(k)$ by equivariance. The point is that for the framed little disks operad, the first three components contain all the information about its homology operad: the generators and relations of $H_*(f\mathcal{D}^{n+1};\mathbb{Q})$ lie in $H_*(f\mathcal{D}^{n+1}(k);\mathbb{Q})$ for $k \leq 3$. This follows from the fact, noticed by Salvatore and Wahl [SW03b], that the operad $H_*(f\mathcal{D}^{n+1};\mathbb{Q})$ is a semidirect product of the homology little disks operad $H_*(\mathcal{D}^{n+1};\mathbb{Q})$ and the rational homology of the group $\mathrm{SO}(n+1)$. They make an explicit computation of the rational homology $H_*(f\mathcal{D}^{n+1};\mathbb{Q})$, see Theorem 2.1.3, which we will use later. The proof is completed by showing that the constructed operad maps respect the operad structures up to homotopy. This implies that there is a homology operad morphism $H_*(f\mathcal{D}^{n+1};\mathbb{Q}) \to H_*(\mathcal{C}^n;\mathbb{Q})$. □

A complete proof along these lines is quite long and ad hoc, as it involves a number of explicit constructions with three disks in the unit disk and explicit homotopies.

In principle, one may give a more direct and, perhaps, shorter proof of this theorem over the rationals, by mapping the generators, explicit in Salvatore-Wahl's theorem, of the operad $H_*(f\mathcal{D}^{n+1};\mathbb{Q})$ to specific cycles in $\mathcal{C}^n$ and checking that the relations are satisfied by showing that the corresponding cycles are homologous in $\mathcal{C}^n$ explicitly. However, the proof outlined earlier seems to be more conceptual, avoids computing the homology of the cacti operad, which might make a good research problem, and hints on the possibility of an operad morphism $f\mathcal{D}^{n+1} \to \mathcal{C}^n$.

It would also be interesting to study the relationship between the kth connected component $\mathcal{C}^n(k)_0$ of the cacti operad and the classifying space of the group of homeomorphisms (or diffeomorphisms) of S^{n+1} with $k+1$ disks removed. In the $n = 1$ case, all these spaces are known to be homotopy equivalent to the framed little disks space $f\mathcal{D}^2(k+1)$ and the moduli space of $(k+1)$-punctured $\mathbb{C}P^1$'s with a real tangent direction at each puncture.

5.2 The cacti action on the sphere space

Let M be an oriented manifold of dimension d. We would like to study continuous k-ary operations on the *(free) sphere space* $SM := M^{S^n} := \mathrm{Map}(S^n, M)$, functorial with respect to M. By passing to singular chains or homology, these operations will induce functorial operations on the chain and homology level, respectively. We will generalize the motivic approach described in Section 2.3 in the $n = 1$ case, following [SV05]. We will use the notation $\mathcal{C} := \mathcal{C}^n$ throughout this section.

Like for $n = 1$, consider the following diagram

$$\mathcal{C}(k) \times (SM)^k \xleftarrow{\rho_{\text{in}}} \mathcal{C}(k)M \xrightarrow{\rho_{\text{out}}} SM \tag{5.1}$$

for each $k \geq 1$, where $\mathcal{C}(k)M$ is the space of triples (c, ϕ, f) with $(c, \phi) \in \mathcal{C}(k)$ and $f : c \to M$ a continuous map from the corresponding cactus to M. The map $\rho_{\text{in}} : \mathcal{C}(k)M \to \mathcal{C}(k) \times (SM)^k$ takes a triple (c, ϕ, f) to $(c, \phi) \in \mathcal{C}(k)$ and the restrictions of f to the k lobes of c. It is an embedding of codimension $d(k-1)$. The map $\rho_{\text{out}} : \mathcal{C}(k)M \to SM$ takes (c, ϕ, f) to the composition of the structure map $\phi : S^n \to c$ with $f : c \to M$. The above diagram defines an operad action in the category $\mathcal{C}orr$ of correspondences, see Section 2.3, on the sphere space SM. Then the proof of the following theorem is no different from the proof of Theorem 2.3.1 for $n = 1$.

Theorem 5.2.1 ([SV05]). 1. *Diagram* (5.1), *considered as a morphism* $\mathcal{C}(k) \times (SM)^k \to SM$ *in* $\mathcal{C}orr$, *defines the structure of a* $\mathcal{C}$*-algebra on the sphere space* SM *in* $\mathcal{C}orr$.

2. *This* $\mathcal{C}$*-algebra structure on the sphere space* SM *in* $\mathcal{C}orr$ *induces an* $h_*(\mathcal{C})$*-algebra structure on the shifted homology* $h_{*+d}(SM)$ *for any multiplicative generalized homology theory* h_* *which supports an orientation of* M.

5.3 The algebraic structure on homology

Combining Theorems 5.1.1 and 5.2.1 we observe the following result.

Corollary 5.3.1. *The shifted rational homology* $H_{*+d}(SM;\mathbb{Q}) := H_*(SM;\mathbb{Q})[d]$ *of the sphere space* $SM = M^{S^n}$ *is an algebra over the operad* $H_*(f\mathcal{D}^{n+1};\mathbb{Q})$.

Taking into account the explicit generators and relations for the operad $H_*(f\mathcal{D}^{n+1};\mathbb{Q})$ given by Theorem 2.1.3, we obtain the following algebraic structure on the rational homology of the sphere space SM.

Corollary 5.3.2. *The (shifted) rational homology* $H_{*+d}(SM;\mathbb{Q})$ *admits the algebraic structure of theorem* 2.1.3. *This includes operations* $\cdot$, $[,]$, $\mathrm{B}_1, \dots$, *and, for* n *odd,* Δ, *satisfying the corresponding identities.*

Remark 10. The dot product, like in the $n = 1$ case, see [CJ02], comes from a *ring* spectrum structure

$$(SM)^{-TM} \wedge (SM)^{-TM} \to (SM)^{-TM},$$

which is constructed using the a standard pinch map $S^n \to S^n \vee S^n$ as a point in $\mathcal{C}^n(2)$ and the diagram

$$SM \times SM \hookleftarrow SM \times_M SM \to SM,$$

which is a specification of (5.1) at that point of $\mathcal{C}^n(2)$ for $k = 2$.

In reality, though, all the unary operations but Δ in the odd n case and $B_{n/2}$ in the even n case vanish on $H_{*+d}(SM;\mathbb{Q})$, as we will see soon. So, perhaps, an analogue of the previous corollary with integral or finite coefficients would bring a lot more information. What we have over $\mathbb{Q}$ can be described using the following definition.

Definition 5.3.1. A BV_{n+1}-*algebra* over a field of zero characteristic is a graded vector space V, along with a dot product ab, a bracket $[a, b]$, and a BV operator Δ of degree n, if n is odd, and an operator $\mathrm{B} = \mathrm{B}_{n/2}$ of degree $2n-1$, if n is even, satisfying the properties (1)–(3) and (6)–(7) of Theorem 2.1.3.

Theorem 5.3.3 ([SV05]). *The shifted rational homology $H_{*+d}(SM;\mathbb{Q})$ of the sphere space $SM = M^{S^n}$ of an oriented manifold M of dimension d has a natural structure of a BV_{n+1}-algebra.*

This theorem follows trivially from Corollary 5.3.2; in other words, in view of Theorem 2.1.3, any $H_*(f\mathcal{D};\mathbb{Q})$-algebra is a BV_{n+1}-algebra. Moreover, one has the following vanishing result.

Proposition 5.3.4 ([SV05]). *The operators B_i, $i \neq n/2$, vanish on $H_{*+d}(SM;\mathbb{Q})$.*

Proof. These operators come from an action of the group $\mathrm{SO}(n+1)$, which is obviously homotopy equivalent to the monoid $f\mathcal{D}^{n+1}(1)$. This monoid maps naturally to $\mathcal{C}^n(1) = \mathrm{Map}(S^n, S^n)$ and acts through it on the sphere space SM. This action is the $k=1$ part of the action from Theorem 5.2.1. Thus, it suffices to perform the following computation of the rational homology of $\mathrm{Map}(S^n, S^n)$. □

Lemma 5.3.5. 1. *We have an isomorphism*

$$H_*(\mathrm{Map}(S^n, S^n);\mathbb{Q}) = \begin{cases} \mathbb{Q}[\Delta, q, q^{-1}] & \text{for } n \text{ odd,} \\ \mathbb{Q}[\mathrm{B}, q, q^{-1}] & \text{for } n \text{ even.} \end{cases}$$

In either case the right-hand side is a graded commutative algebra on one generator q of degree zero and one odd generator Δ of degree n for n odd or one odd generator B of degree $2n-1$ for n even, localized at $q = 0$.

2. *Under the map $\mathrm{SO}(n+1) \to \mathrm{Map}(S^n, S^n)$, the elements Δ and $\mathrm{B}_{n/2} \in H_*(\mathrm{SO}(n+1);\mathbb{Q})$ map to the above Δ and B, while the other B_i's map to zero.*

Idea of proof of Lemma, after Sullivan. The proof uses a method of rational homotopy theory, which finds the (commutative, minimal) DGA representing the structure group of a fiber bundle, given a DGA representing a generic base and a DGA representing the fiber. We will compute the homology of $\mathrm{Map}(S^4, S^4)$ as an example. Here we are talking about an S^4-bundle over a base represented by a DGA A. The DGA of the fiber is $\mathbb{Q}[u, v]$ with degrees $|u| = 4$ and $|v| = 7$ and differential $du = 0$ and $dv = u^2$. Then the DGA model of the total space may be obtained as $A[u, v]$ with a differential determined by $du = a$ for some $a \in A$

and $dv = u^2 + pu + q$ for some $p, q \in A$. Of course, we should have $d^2 = 0$, so $d^2v = 0$ yields $2ua + pa + udp + dq = 0$ or $2a + dp = 0$ and $pa + dq = 0$. If we set $u' := u - p/2$, we will have $du' = 0$ and $dv = (u')^2 + q - p^2/4$. Thus, the total space DGA is now universally presented as $A[u', v]$ with the above differential. Note that for $q' := q - p^2/4$ we have $dq' = 0$. Therefore, an S^4-bundle is determined in rational homotopy theory by a characteristic class $q' \in A$ of degree eight and the classifying space must rationally be $K(\mathbb{Q}, 8)$. By transgression, the structure group must rationally be $K(\mathbb{Q}, 7) \sim_{\mathbb{Q}} S^7$. One needs some extra work to see that a homology generator of this space comes from the generator B of $H_7(\mathrm{SO}(5); \mathbb{Q})$. □

5.4 Sphere spaces and Hochschild homology

Here we will announce a result of the second author, which provides an approximation of the sphere space $SX = X^{S^n}$ via configuration spaces with labels and implies a computation of the homology of SX as the Hochschild homology of the n-algebra $C_*(\Omega^n X)$, where $\Omega^n X = \mathrm{Map}_*(S^n, X)$ is the based n-fold loop space of a space X. These results directly generalize the results of Burghelea and Fiedorowicz [BF86] for loop spaces ($n = 1$). The first computation of the homology of SX as an n-algebra Hochschild homology was done by Po Hu [Hu04], who generalized the Chen-Getzler-Jones-Petrack-Segal method. Unfortunately, we can only speculate that our version of the n-algebra Hochschild homology is isomorphic to the one used by Po Hu.

Theorem 5.4.1. *The topological Hochschild complex (see below) of the based n-fold loop space $\Omega^n X$, considered as an $f\mathcal{D}^n$-algebra is homeomorphic to the space $X_0^{S^n}$ of maps $S^n \to X$ passing through the basepoint of a space X.*

Proof. Let us first define the topological Hochschild complex; when this is done, the proof will be fairly straightforward.

From this point on, we need to consider a version of the framed little n-disks operad, namely the colored operad of labeled, framed "little" (closed) disks inside the standard "large" (open) disk of a varying radius r, $0 < r \leq \pi$, centered at the north pole on the standard unit sphere $S^n \subset \mathbb{R}^{n+1}$. A disk on the sphere is understood as a disk (ball) in the standard metric on the sphere induced from the ambient Euclidean $\mathbb{R}^{n+1}$. The colors are given by the radii r of the disks. A frame in a little disk is a positively oriented orthonormal frame in the tangent space to the sphere at the center of the disk. The frame in the standard spherical disk around the north pole is the part $e_1, \dots, e_n$ of the standard orthonormal basis of $\mathbb{R}^{n+1}$. The operad composition is given by rotating a given large disk of radius r with a few little disks inside it to the place of a prescribed little disk of radius r in another large spherical disk to fit the frames and gluing the first large disk into the prescribed given little disk inside the second large disk. We will denote the resulting framed spherical little disks operad $f\mathcal{D}^n_s$.

Consider a collection $f\mathcal{D}S^n = \{f\mathcal{D}S^n(k) \mid k \geq 1\}$ of similar spaces oflabeled, framed little disks around any point on the whole unit sphere S^n. This collection is a (right) module over the colored operad $f\mathcal{D}^n_s$ via similar gluing after an appropriate rotation, as in the definition of the colored operad structure on $f\mathcal{D}^n_s$ above.

Suppose a space A is an algebra over the (colored) operad $\mathcal{O}$ in the category of topological spaces and M an $\mathcal{O}$-module. Then we can form the *tensor product of M and A over $\mathcal{O}$* as follows:

$$M \otimes_{\mathcal{O}} A := \coprod_k M(k) \times_{\Sigma_k} A^k / \sim,$$

where $\sim$ denotes the equivalence relation which may be described roughly as $(mo, a) \sim (m, oa)$ for $m \in M$, $o \in \mathcal{O}$, and $a \in A$. In other words, $M \otimes_{\mathcal{O}} A$ is the coequalizer of the diagram $M \times_\Sigma \mathcal{O} \times_\Sigma \times A \rightrightarrows M \times_\Sigma A$.

Now let A be an algebra over the colored operad $f\mathcal{D}^n_s$, for example, a Moore-type version of the based n-fold loop space $\Omega^n X$ of a pointed space X. This Moore-type version $\Omega^n_s X$ may be defined as the space of continuous maps from a closed disk of radius r, $0 < r \leq \pi$, centered at the north pole of the sphere S^n, mapping the boundary of the disk to the basepoint in X. Then, by definition, the *topological Hochschild homology of the $f\mathcal{D}^n_s$-algebra A* is the tensor product $f\mathcal{D}S^n \otimes_{f\mathcal{D}^n_s} A$.

To prove the theorem, we thereby need to present a homeomorphism

$$f\mathcal{D}S^n \otimes_{f\mathcal{D}^n_s} \Omega^n_s X \cong X_0^{S^n}.$$

We define it by assigning to a configuration of k framed little disks on S^n and a given collection of k maps of the standard disks around the north pole on the sphere S^n of matching radii to X (representing k points in $\Omega^n_s X$) to a map from the sphere S^n to X by using the k maps on the little disks (prepended by appropriate rotations, as above) and extending these maps by a constant to a map $S^n \to X$. This gives a point in $X_0^{S^n}$.

The inverse of this homeomorphism is given by taking, for a given map $S^n \to X$ passing through the basepoint, a point on $p \in S^n$ mapping to the basepoint and thinking of its complement $S^n \setminus \{p\}$ as one little disk of radius π on S^n, providing it with any framing, which determines a map from S^n minus the south pole, regarded as the disk at the north pole of radius π, to $S^n \setminus \{p\}$ mapping the south pole to p, and thus getting an element of $\Omega^n_s X$. This combines into a map $X_0^{S^n} \to f\mathcal{D}S^n(1) \times \Omega^n_s X$, which we map into the tensor product $f\mathcal{D}S^n \otimes_{f\mathcal{D}^n_s} \Omega^n_s X$ naturally. □

The relation between the free sphere space X^{S^n} and the "semifree" one $X_0^{S^n}$ is amazingly simple.

Proposition 5.4.2. *For a path connected space X having numerable category, in particular, a connected CW complex, the inclusion $X_0^{S^n} \subset X^{S^n}$ is a homotopy equivalence.*

Proof. Consider the map $p : X^{S^n} \to X$ evaluating a map $S^n \to X$ at the north pole. One checks that it is a fibration by verifying the homotopy lifting property directly. There is a similar fibration $X_0^{S^n} \to X$ and a commutative square

$$\begin{array}{ccc} X_0^{S^n} & \longrightarrow & X^{S^n} \\ \downarrow & & \downarrow \\ X & \xrightarrow{\text{id}} & X. \end{array}$$

It suffices to show that the map induced on the fiber is a homotopy equivalence. Indeed the map induced between the fibers over the basepoint is just the identity map. □

We can pass to singular cochains C_* and homology H_* and obtain the following corollary.

Corollary 5.4.3. 1. *The Hochschild chain complex $C_*^{(n)}(C_*\Omega_s^n X, C_*\Omega_s^n X)$ of the $C_*f\mathcal{D}_s^n$-algebra $C_*\Omega_s^n X$ is homotopy equivalent to the singular cochain complex $C_*X_0^{S^n}$.*

2. *The Hochschild homology $H_*^{(n)}(C_*\Omega_s^n X, C_*\Omega_s^n X)$ of the $C_*f\mathcal{D}_s^n$-algebra $C_*\Omega_s^n X$ is isomorphic to the homology $H_*(X_0^{S^n};\mathbb{Z})$.*

If X is path connected and has numerable category, one can replace $X_0^{S^n}$ by the sphere space X^{S^n} in the above.

Here the *Hochschild chain complex* $C_*^{(n)}(A, A)$ of a $C_*f\mathcal{D}_s^n$-algebra A is understood as $C_*f\mathcal{D}S^n \otimes_{C_*f\mathcal{D}_s^n} A$ and the *Hochschild homology* $H_*^{(n)}(A, A)$ as the homology of this complex. Motivation for this definition may be the $n = 1$ case, in which $C_*f\mathcal{D}S^1$ and $C_*f\mathcal{D}_s^1$ must be replaced by the cofibrant models $C_*\overline{C}(S^1)$ and $C_*\overline{C}(D^1)$, where $\overline{C}$ stands for the Fulton-MacPherson compactification and C_* for cellular chains. After these changes, the tensor product becomes a configuration space with summable labels, see [BF86, Sal01], and we get the usual Hochschild complex.

As we mentioned earlier, the first result of this type was obtained by Po Hu [Hu04], who used a different notion of the Hochschild homology. It will be very interesting to find a relationship between them. Hopefully, our result is a geometric incarnation of that of Po Hu.

Brane topology and Kontsevich's Hochschild cohomology conjecture

We would like to end the discussion with the following table of analogies between algebra and topology. Since the discussion is highly speculative, we will freely confuse the notion of an n-algebra, which is an algebra over the operad $H_*(\mathcal{D}^n;\mathbb{Z})$, see Section 2.1.4, with that of an algebra over the chain operad $C_*(\mathcal{D}^n;\mathbb{Z})$ or the topological operad $\mathcal{D}^n$ and an n_∞-algebra. Recall that a 1-algebra is the same as an associative algebra and a 2-algebra is the same as a G-algebra.

Algebra	Topology
a 1-algebra A	a loop space ΩM is a 1-algebra
$H^*(A,A)$ is a 2-algebra by Gerstenhaber	$H_{*+d}(LM)$ is a 2-algebra by Chas-Sullivan
	$H_{*+d}(C_*\Omega M, C_*\Omega M)$, a graded abelian group isomorphic to $H_{*+d}(LM)$ by Burghelea-Fiedorowicz, is a 2-algebra, via Poincaré duality and the fact that $C_*\Omega M$ is a 1-algebra
an n-algebra A	a loop space $\Omega^n M$ is an n-algebra
$H^*_{(n)}(A,A)$ is an $(n+1)$-algebra by Kontsevich	$H_{*+d}(M^{S^n})$ is an $(n+1)$-algebra by Sullivan-AV
	$H^{(n)}_{*+d}(C_*\Omega^n M, C_*\Omega^n M)$, a graded abelian group isomorphic to $H_{*+d}(M^{S^n})$ by Po Hu, is an $(n+1)$-algebra, via Poincaré duality and the fact that $C_*\Omega^n M$ is an n-algebra

It would be very interesting to find a relation of our notion of Hochschild n-algebra homology to an algebraic one, see, for example, [Kon99, Tam00, HKV01], and check if the isomorphisms in the right column of the table respected the n-algebra structures. One may think of such a statement as a topological incarnation of Kontsevich's conjecture on Hochschild cohomology, proven in [Tam00, HKV01]: the Hochschild cochain complex of an n-algebra is naturally an $(n+1)_\infty$-algebra.

Bibliography

[Abr96] L. Abrams, *Two-dimensional topological quantum field theories and Frobenius algebras*, J. Knot Theory Ramifications **5** (1996), no. 5, 569–587. MR MR1414088 (97j:81292)

[Ada78] J. F. Adams, *Infinite loop spaces*, Princeton University Press, Princeton, NJ, 1978.

[AS04] A. Abbondandolo and M. Schwarz, *On the Floer homology of cotangent bundles*, Preprint, August 2004, `math.SG/0408280`.

[Ati61] M. F. Atiyah, *Thom complexes*, Proc. London Math. Soc. **3** (1961), no. 11, 291–310.

[Ati88] M. Atiyah, *Topological quantum field theories*, Inst. Hautes Études Sci. Publ. Math. (1988), no. 68, 175–186 (1989). MR MR1001453 (90e:57059)

[BC94] M. Betz and R. L. Cohen, *Moduli spaces of graphs and cohomology operations*, Turkish Journal of Math. **18** (1994), 23–41.

[BCR04] N. A. Baas, R. L. Cohen, and A. Ramirez, *The topology of the category of open and closed strings*, Preprint, November 2004, `math.AT/0411080`.

[Beh97] K. Behrend, *Gromov-Witten invariants in algebraic geometry*, Invent. Math. **127** (1997), no. 3, 601–617, `alg-geom/9601011`. MR MR1431140 (98i:14015)

[BF86] D. Burghelea and Z. Fiedorowicz, *Cyclic homology and algebraic K-theory of spaces. II*, Topology **25** (1986), no. 3, 303–317. MR MR842427 (88i:18009b)

[BG75] J. C. Becker and D. Gottlieb, *The transfer maps and fibre bundles*, Topology **14** (1975), 1–12.

[BHM93] M. Bokstedt, W.-C. Hsiang, and I. Madsen, *The cyclotomic trace and algebraic K-theory of spaces*, Invent. Math. **111** (1993), 465–540.

[BV73] J. M. Boardman and R. M. Vogt, *Homotopy invariant algebraic structures on topological spaces*, Lecture Notes in Math., vol. 347, Springer-Verlag, 1973.

[CG04] R. L. Cohen and V. Godin, *A polarized view of string topology*, Topology, geometry and quantum field theory, London Math. Soc. Lecture Notes, vol. 308, Cambridge Univ. Press, Cambridge, 2004, `math.AT/0303003`, pp. 127–154.

[Cha01] M. Chas, *Combinatorial Lie bialgebras of curves on surfaces*, Tech. report, SUNY at Stony Brook, May 2001, `math.GT/0105178`. To appear in Topology.

[Cha03] D. Chataur, *A bordism approach to string topology*, Preprint, CRM, Barcelona, 2003, `arXiv:math.AT/0306080`.

[CJ02] R. L. Cohen and J. D. S. Jones, *A homotopy theoretic realization of string topology*, Math. Ann. **324** (2002), no. 4, 773–798, `math.GT/0107187`. MR 1 942 249

[CJY03] R. L. Cohen, J. D. S. Jones, and J. Yan, *The loop homology algebra of spheres and projective spaces*, Progr. Math., vol. 215, Birkhäuser, Basel, 2003, pp. 77–92.

[CK00] A. Connes and D. Kreimer, *Renormalization in quantum field theory and the Riemann-Hilbert problem. I. The Hopf algebra structure of graphs and the main theorem*, Comm. Math. Phys. **210** (2000), no. 1, 249–273, `hep-th/9912092`. MR 2002f:81070

[CKS05] R. L. Cohen, J. Klein, and D. Sullivan, *On the homotopy invariance of string topology*, In preparation, March 2005.

[CN05] R. L. Cohen and P. Norbury, *Morse field theory*, In preparation, March 2005.

[Coh76] F. R. Cohen, *The homology of $\mathcal{C}_{n+1}$-spaces, $n \geq 0$*, The homology of iterated loop spaces, Lecture Notes in Math., vol. 533, Springer-Verlag, 1976, pp. 207–351.

[Coh04a] R. L. Cohen, *Morse theory, graphs, and string topology*, Preprint, November 2004, `math.GT/0411272`. To appear in Proc. SMS/NATO Adv. study inst. on Morse theoretic methods in nonlinear analysis and symplectic topology, Kluwer press (2005).

[Coh04b] ———, *Multiplicative properties of Atiyah duality*, Homology, Homotopy, and its Applications **6** (2004), no. 1, 269–281.

[Coh05] ———, *String topology and Morse theory on the loop space*, In preparation, March 2005.

[CS99] M. Chas and D. Sullivan, *String topology*, Preprint, CUNY, November 1999, `math.GT/9911159`. To appear in Ann. of Math.

[DW90] R. Dijkgraaf and E. Witten, *Topological gauge theories and group cohomology*, Comm. Math. Phys. **129** (1990), no. 2, 393–429. MR MR1048699 (91g:81133)

[EKMM97] A. D. Elmendorf, I. Kriz, M. A. Mandell, and J. P. May, *Rings, modules, and algebras in stable homotopy theory*, Math. Surveys and Monographs, vol. 47, Amer. Math. Soc., Providence, 1997.

[Fie92] Z. Fiedorowicz, *The symmetric bar construction*, Preprint, Ohio State University, 1992, `http://www.math.ohio-state.edu/~fiedorow/`.

[Fie98] ______, *Constructions of E_n operads*, Preprint, Ohio State University, August 1998, `math.AT/9808089`.

[FMT02] Y. Felix, L. Menichi, and J.-C. Thomas, *Duality in Gerstenhaber algebras*, Preprint, November 2002, `math.AT/0211229`.

[Fuk93] K. Fukaya, *Morse homotopy, A_∞-category, and Floer homologies*, Proc. Garc. workshop on geometry and topology (Seoul), Lecture Notes Ser., vol. 18, Seoul National Univ., 1993, pp. 1–102.

[Ger68] M. Gerstenhaber, *On the deformation of rings and algebras. III*, Ann. of Math. (2) **88** (1968), 1–34. MR MR0240167 (39 #1521)

[Get94] E. Getzler, *Batalin-Vilkovisky algebras and two-dimensional topological field theories*, Comm. Math. Phys. **159** (1994), no. 2, 265–285, `hep-th/9212043`.

[God04] V. Godin, Ph.D. thesis, Stanford University, Stanford, CA, 2004.

[Gol86] W. M. Goldman, *Invariant functions on Lie groups and Hamiltonian flows of surface group representations*, Invent. Math. **85** (1986), no. 2, 263–302.

[Goo85] T. Goodwillie, *Cyclic homology, derivations, and the free loop space*, Topology **24** (1985), 187–215.

[Gru05] K. Gruher, Ph.D. thesis, Stanford University. In preparation, March 2005.

[Har85] J. L. Harer, *Stability of the homology of the mapping class groups of orientable surfaces*, Ann. of Math. **121** (1985), 215–249.

[Har04] E. Harrelson, *On the homology of open-closed string theory*, Preprint, December 2004, `math.AT/0412249`.

[HKV01] P. Hu, I. Kriz, and A. A. Voronov, *On Kontsevich's Hochschild cohomology conjecture*, Preprint, University of Michigan, September 2001, `math.AT/0309369`. To appear in Compos. Math.

[HSS00] M. Hovey, B. Shipley, and J. Smith, *Symmetric spectra*, J. Amer. Math. Soc. **13** (2000), 149–208.

[Hu04] P. Hu, *Higher string topology on general spaces*, Preprint, Wayne State University, January 2004, `math.AT/0401081`.

[Hua97] Y.-Z. Huang, *Two-dimensional conformal geometry and vertex operator algebras*, Birkhäuser Boston, Boston, MA, 1997. MR 98i:17037

[Jon87] J. D. S. Jones, *Cyclic homology and equivariant homology*, Inv. Math. **87** (1987), 403–423.

[Kle99] J. Klein, *Poincaré duality embeddings and fiberwise homotopy theory*, Topology **38** (1999), 597–620.

[Kle03] ———, *Fiber products, Poincaré duality, and A_∞-ring specta*, Preprint, June 2003, `math.AT/0306350`.

[Kli82] W. Klingenberg, *Riemannian geometry*, Math. Surveys and Monographs, vol. 47, deGruyter, 1982.

[KM94] M. Kontsevich and Yu. I. Manin, *Gromov-Witten classes, quantum cohomology, and enumerative geometry*, Comm. Math. Phys. **164** (1994), 525–562, `hep-th/9402147`.

[Kon92] M. Kontsevich, *Intersection theory on the moduli space of curves and the matrix Airy function*, Comm. Math. Phys. **147** (1992), no. 1, 1–23. MR 93e:32027

[Kon99] ———, *Operads and motives in deformation quantization*, Lett. Math. Phys. **48** (1999), 35–72.

[KS00] M. Kontsevich and Y. Soibelman, *Deformations of algebras over operads and the Deligne conjecture*, Conférence Moshé Flato 1999, Vol. I (Dijon), Kluwer Acad. Publ., Dordrecht, 2000, `math.QA/0001151`, pp. 255–307. MR 1 805 894

[KS03] S. Kallel and P. Salvatore, *Rational maps and string topology*, Preprint, Université Lille I, 2003, `math.AT/0309038`.

[KSV96] T. Kimura, J. Stasheff, and A. A. Voronov, *Homology of moduli spaces of curves and commutative homotopy algebras*, The Gelfand Mathematics Seminars, 1993–1995 (I. Gelfand, J. Lepowsky, and M. M. Smirnov, eds.), Birkhäuser Boston, 1996, pp. 151–170.

[Lal04] F. Lalonde, *A field theory for symplectic fibrations over surfaces*, Geom. Topol. **8** (2004), 1189–1226 (electronic), `math.SG/0309335`. MR MR2087081

[LT98] J. Li and G. Tian, *Virtual moduli cycles and Gromov-Witten invariants of algebraic varieties*, J. Amer. Math. Soc. **11** (1998), no. 1, 119–174. MR MR1467172 (99d:14011)

[Man99] Y. I. Manin, *Frobenius manifolds, quantum cohomology, and moduli spaces*, American Mathematical Society Colloquium Publications, vol. 47, American Mathematical Society, Providence, RI, 1999. MR MR1702284 (2001g:53156)

[May96] J. P. May, *Operads, algebras and modules*, Operads: Proceedings of Renaissance Conferences (J.-L. Loday, J. Stasheff, and A. A. Voronov, eds.), Amer. Math. Soc., 1997, pp. 15–31.

[Mer03] S. A. Merkulov, *A de Rham model for string topology*, Preprint, Stockholm University, 2003, `arXiv:math.AT/0309038`.

[ML65] S. Mac Lane, *Categorical algebra*, Bull. Amer. Math. Soc. **71** (1965), 40–106. MR 30 #2053

[MS02] J. E. McClure and J. H. Smith, *A solution of Deligne's Hochschild cohomology conjecture*, Recent progress in homotopy theory (Baltimore, MD, 2000), Amer. Math. Soc., Providence, RI, 2002, `math.QA/9910126`, pp. 153–193. MR 1 890 736

[MSS02] M. Markl, S. Shnider, and J. Stasheff, *Operads in algebra, topology and physics*, Mathematical Surveys and Monographs, vol. 96, American Mathematical Society, Providence, RI, 2002. MR 1 898 414

[MV03] M. Markl and A. A. Voronov, *PROPped up graph cohomology*, Preprint M/03/47, IHES, Bures-sur-Yvette, France, July 2003, `math.QA/0307081`. Submitted to Acta Math.

[MW02] I. Madsen and M. S. Weiss, *The stable moduli space of Riemann surfaces: Mumford's conjecture*, Preprint, Aarhus University, December 2002, `math.AT/0212321`.

[Pen87] R. C. Penner, *The decorated Teichmüller space of punctured surfaces*, Comm. Math. Phys. **113** (1987), no. 2, 299–339. MR 89h:32044

[Ram05] A. Ramirez, Ph.D. thesis, Stanford University. In preparation, March 2005.

[Sal01] P. Salvatore, *Configuration spaces with summable labels*, Cohomological methods in homotopy theory (Bellaterra, 1998), Progr. Math., vol. 196, Birkhäuser, Basel, 2001, `math.AT/9907073`, pp. 375–395. MR 2002f:55039

[Sch93] A. Schwarz, *Geometry of Batalin-Vilkovisky quantization*, Comm. Math. Phys. **155** (1993), no. 2, 249–260. MR MR1230027 (95f:81095)

[Sch05] M. Schwarz, Proc. SMS/NATO Adv. study inst. on Morse theoretic methods in nonlinear analysis and symplectic topology, Kluwer Press, 2005. To appear.

[Seg88] G. Segal, *Elliptic cohomology (after Landweber-Stong, Ochanine, Witten, and others)*, Astérisque (1988), no. 161-162, Exp. No. 695, 4, 187–201 (1989), Séminaire Bourbaki, Vol. 1987/88. MR MR992209 (91b:55005)

[Seg99] ———, *Topological field theory ('Stanford Notes')*, `http://www.cgtp.duke.edu/ITP99/segal/`, 1999.

[Seg01] ———, *Topological structures in string theory*, Phil. Trans. R. Soc. Lond. A **359** (2001), 1389–1398.

[Seg04] ———, *The definition of conformal field theory*, Topology, geometry and quantum field theory, London Math. Soc. Lecture Notes, vol. 308, Cambridge Univ. Press, Cambridge, 2004, pp. 421–577. MR MR2079383

[Sta63] J. Stasheff, *On the homotopy associativity of H-spaces, I*, Trans. Amer. Math. Soc. **108** (1963), 275–292.

[Str84] K. Strebel, *Quadratic differentials*, Springer-Verlag, Berlin, 1984.

[Sul04] D. Sullivan, *Open and closed string field theory interpreted in classical algebraic topology*, Topology, geometry and quantum field theory, London Math. Soc. Lecture Notes, vol. 308, Cambridge Univ. Press, Cambridge, 2004, `math.QA/0302332`, pp. 344–357.

[SV05] D. Sullivan and A. A. Voronov, *Brane topology*, Preprint, University of Minnesota, In preparation, March 2005.

[SW03a] D. Salamon and J. Weber, *Floer homology and the heat flow*, Preprint, April 2003, `math.SG/0304383`.

[SW03b] P. Salvatore and N. Wahl, *Framed discs operads and Batalin-Vilkovisky algebras*, Q. J. Math. **54** (2003), no. 2, 213–231, `math.AT/0106242`.

[Tam00] D. E. Tamarkin, *Deformation complex of a d-algebra is a (d + 1)-algebra*, Preprint, Harvard University, October 2000, `math.QA/0010072`.

[Til97] U. Tillmann, *On the homotopy of the stable mapping class group*, Invent. Math. **130** (1997), no. 2, 257–275.

[Vit96] C. Viterbo, *Functors and computations in Floer homology with applications, Part II*, Preprint, 1996.

[Vor01] A. A. Voronov, *Notes on universal algebra*, Graphs and patterns in mathematics and theoretical physics, Proc. Sympos. Pure Math., vol. 73, Amer. Math. Soc., Providence, RI, 2005, pp. 81–103, `math.QA/0111009`.

[Web02] J. Weber, *Perturbed closed geodesics are periodic orbits: Index and transversality*, Math. Zeit. **241** (2002), 45–81.

[Wit91] E. Witten, *Introduction to cohomological field theories*, Internat. J. Modern Phys. A **6** (1991), no. 16, 2775–2792, Topological methods in quantum field theory (Trieste, 1990). MR MR1117747 (92h:58035)

[Wol83] S. Wolpert, *On the symplectic geometry of deformations of hyperbolic surfaces*, Ann. of Math. **117** (1983), 207–234.

Part II

An Algebraic Model for Mod 2 Topological Cyclic Homology

Kathryn Hess

Preface

Bökstedt, Hsiang and Madsen introduced *topological cyclic homology (TC)* as a topological version of Connes' cyclic homology in [BHM]. The topological cyclic homology of a space X at a prime p, denoted $TC(X;p)$, is a spectrum that is the target of the *cyclotomic trace map*

$$Trc\colon A(X) \longrightarrow TC(X;p),$$

the source of which is Waldhausen's algebraic K-theory spectrum of X, to which $TC(X;p)$ provides a useful approximation. The cyclotomic trace map is analogous to the Dennis trace map $K_*(A) \longrightarrow HH_*(A)$. Very little is known about the $TC(X;p)$ when X is not a singleton.

Waldhausen's algebraic K-theory itself approximates the smooth and topological Whitehead spectra $Wh^d(X)$ and $Wh^t(X)$. There are natural cofiber sequences of spectra

$$\Sigma^\infty X_+ \xrightarrow{\eta_X} A(X) \longrightarrow Wh^d(X)$$

and

$$A(*) \wedge X_+ \xrightarrow{a_X^A} A(X) \longrightarrow Wh^t(X)$$

respectively, where η_X is the unit map and a_X^A is the A-theory assembly map [W1, 3.3.1]. Here, Y_+ denotes the space Y with an extra basepoint, and $\Sigma^\infty Y$ denotes the suspension spectrum of Y.

By Waldhausen's stable parametrized h-cobordism theorem [W2], there are homotopy equivalences

$$\mathcal{H}^d(M) \simeq \Omega\Omega^\infty Wh^d(M)$$

and

$$\mathcal{H}^t(M) \simeq \Omega\Omega^\infty Wh^t(M)$$

when M is a smooth, respectively topological, compact manifold. Here $\mathcal{H}^d(M)$ is the stable smooth h-cobordism space of M, which in a stable range carries information about the homotopy type of the topological group $\mathrm{Diff}(M)$ of self-diffeomorphisms of M. Likewise $\mathcal{H}^t(M)$ is the stable topological h-cobordism space of M, which in a stable range carries information about the topological group Homeo(M) of self-homeomorphisms of M.

Any information we may obtain about the topological cyclic homology of a smooth or topological manifold will therefore give us some indication of the nature of its stable h-cobordism space.

For further explanation of the role of trace maps in K-theory and (topological) cyclic and Hochschild homology, as well as about Whitehead spectra, we refer the reader to [Be] and [R].

The goal of these notes, as well as of the minicourse upon which they are based, is to construct a cochain complex $tc^*(X)$ such that $\mathrm{H}^*(tc^*(X) \otimes \mathbb{F}_p)$ is isomorphic to the mod p spectrum cohomology of $TC(X;p)$. For reasons of ease of notation and computation, we will limit ourselves to $p = 2$ in this article.

There are several equivalent definitions of $TC(X;p)$. The definition that is best suited to algebraic modeling can be stated as follows [BHM]. Let $\mathcal{L}X$ be the free loop space on X, i.e., the space of unbased maps from the circle S^1 into X, which admits a natural S^1-action, by rotation of loops. Let $\mathcal{L}X_{hS^1} = ES^1 \underset{S^1}{\times} \mathcal{L}X$ denote the homotopy orbit space of this action. Let $\lambda^p : \mathcal{L}X \longrightarrow \mathcal{L}X$ denote the pth-power map, defined by $\lambda^p(\ell)(z) = \ell(z^p)$ for all $\ell \in \mathcal{L}X$ and all $z \in S^1$. There is a homotopy pullback of spectra

$$\begin{array}{ccc} TC(X;p) & \longrightarrow & \Sigma^\infty \mathcal{L}X_+ \\ \downarrow & & \downarrow{\scriptstyle \Sigma^\infty(Id-\lambda^p)} \\ \Sigma^\infty\big(\Sigma(\mathcal{L}X_{hS^1})_+\big) & \xrightarrow{trf_{S^1}} & \Sigma^\infty \mathcal{L}X_+ \end{array}$$

where trf_{S^1} is the S^1-transfer map associated to the principal S^1-bundle

$$ES^1 \times \mathcal{L}X \longrightarrow \mathcal{L}X_{hS^1}.$$

It is therefore clear that $TC(X;p)$ is the homotopy fiber of the composition

$$\Sigma^\infty \mathcal{L}X_+ \xrightarrow{\Sigma^\infty(Id-\lambda^p)} \Sigma^\infty \mathcal{L}X_+ \xrightarrow{\iota} \mathrm{hocofib}(trf_{S^1})\ .$$

Motivated by this characterization of $TC(X;p)$, we apply the following method to constructing $tc^*(X)$. We first define an associative cochain algebra $fls^*(X)$ together with a cochain map

$$\Upsilon : fls^*(X) \longrightarrow CU^*\mathcal{L}X$$

inducing an isomorphism of algebras in cohomology, where CU^* denotes the reduced cubical cochains. We then twist together $fls^*(X)$ and $\mathrm{H}^*(BS^1)$, obtaining a new cochain complex $hos^*(X)$ that fits into a commuting diagram

$$\begin{array}{ccc} hos^*(X) & \xrightarrow{\pi} & fls^*(X) \\ \downarrow{\scriptstyle \overline{\Upsilon}} & & \downarrow{\scriptstyle \Upsilon} \\ CU^*(\mathcal{L}X_{hS^1}) & \xrightarrow{CU^*c} & CU^*(\mathcal{L}X) \end{array}$$

where π is the projection map, $c : \mathcal{L}X \longrightarrow \mathcal{L}X_{hS^1}$ is the map induced by the inclusion $\mathcal{L}X \longrightarrow ES^1 \times \mathcal{L}X$ and $\overline{\Upsilon}$ induces an isomorphism in cohomology. The projection map $\pi : hos^*(X) \longrightarrow fls^*(X)$ is then a model for the inclusion

$$\Sigma^\infty \mathcal{L}X_+ \xrightarrow{\iota} \text{hocofib}(trf_{S^1}) .$$

Finally, we define a cochain map $\text{\it\u{l}}^p : fls^*(X) \longrightarrow fls^*(X)$ such that $\Upsilon \circ \text{\it\u{l}}^p$ and $CU^*\lambda^p \circ \Upsilon$ are chain homotopic. Thus $Id - \text{\it\u{l}}^p$ is a model for $\Sigma^\infty(Id - \lambda^p)$. As explained carefully and in detail in [HR], if we set $tc^*(X)$ equal to the mapping cone of the composition $(Id - \text{\it\u{l}}^p)\pi$, then $\mathrm{H}^*(tc^*(X) \otimes \mathbb{F}_p) \cong \mathrm{H}^*(TC(X;p);\mathbb{F}_p)$, as desired. Here we remark only that there is a Thom isomorphism involved in the identification of the cohomology of the homotopy cofiber of the S^1-transfer map and the cohomology of the S^1-homotopy orbits of the free loop space. Furthermore, the fact that the projection map π is a model for the inclusion ι requires an analysis of these transfers and isomorphisms

The article is organized as follows. We begin in chapter 1 by reminding the reader of certain algebraic and topological notions and constructions. In chapter 2 we study free loop spaces and their algebraic models, beginning by defining a simplicial set that models $\mathcal{L}X$, which we then apply to constructing $fls^*(X)$ via a refined version of methods from [DH1]-[DH4]. Chapter 3 is devoted to the study of homotopy orbit spaces of circle actions. We first treat the general case, twisting together $\mathrm{H}^*(BS^1)$ and CU^*Y to obtain a large but attractive cochain complex $HOS^*(Y)$ for calculating $\mathrm{H}^*(Y_{hS^1})$, when Y is any S^1-space. Specializing to the case $Y = \mathcal{L}X$, we show how to twist together $\mathrm{H}^*(BS^1)$ and $fls^*(X)$ to build $hos^*(X)$, so that we obtain a complex equivalent to $HOS^*(\mathcal{L}X)$. Finally, in chapter 4 we define and study our model for the pth-power map (for $p = 2$), then apply it, together with the results of the preceding chapters, to the construction of $tc^*(X)$. We conclude by applying our model to the calculation of the mod 2 spectrum cohomology of $TC(S^{2n+1}; 2)$.

Remark. In these notes, complete proofs are provided only of those results that have not yet appeared elsewhere and that are due to the author. Furthermore, some results that have yet to be published are not proved completely here, if the complete proof is excessively technical. We hope in such cases to have provided enough detail to convince the reader of the truth of the statement. The reader who is curious about the details is refered to articles that should appear soon.

The author would like to thank David Chataur, John Rognes and Jérôme Scherer for their helpful comments on earlier versions of these notes. Warm thanks are also due to David Chataur, José-Luis Rodrigues and Jérôme Scherer for their splendid organization of the Almería summer school on string topology.

Chapter 1

Preliminaries

We begin here by recalling certain elementary definitions and constructions and fixing our basic notation and terminology. We then remind the reader of the construction of the canonical, enriched Adams-Hilton model of a simplicial set, which is the input data for our free loop space model. We conclude this chapter with a description of our general method for constructing algebraic models of fiber squares, which we then apply in chapter 2 to building our free loop space model.

1.1 Elementary definitions, terminology and notation

Throughout this paper we work over $\mathbb{Z}$, the ring of integers, unless stated otherwise.

Given chain complexes (V, d) and (W, d), the notation $f : (V, d) \xrightarrow{\simeq} (W, d)$ indicates that f induces an isomorphism in homology. In this case we refer to f as a *quasi-isomorphism*.

If $V = \bigoplus_{i \in \mathbb{Z}} V_i$ is a graded module, then $s^{-1}V$ and sV denote the graded modules with, respectively, $(s^{-1}V)_i \cong V_{i+1}$ and $(sV)_i \cong V_{i-1}$. Given a homogeneous element v in V, we write $s^{-1}v$ and sv for the corresponding elements of $s^{-1}V$ and sV. If the gradings are written as upper indices, i.e., $V = \bigoplus_{i \in \mathbb{Z}} V^i$, then $(s^{-1}V)^i \cong V^{i-1}$ and $(sV)^i \cong V^{i+1}$.

Dualization is indicated throughout the paper by a $\sharp$ as superscript. The degree of an element x in a graded module is denoted $|x|$, unless it is used as an exponent, in which case the bars may be dropped.

A graded $\mathbb{Z}$-module $V = \bigoplus_{i \in \mathbb{Z}} V_i$ is *connected* if $V_{<0} = 0$ and $V_0 \cong \mathbb{Z}$. It is *simply connected* if, in addition, $V_1 = 0$. We write V_+ for $V_{>0}$. Let V be a positively-graded, free $\mathbb{Z}$-module. The free associative algebra on V is denoted TV, i.e.,

$$TV \cong \mathbb{Z} \oplus V \oplus (V \otimes V) \oplus (V \otimes V \otimes V) \oplus \cdots .$$

A typical basis element of TV is denoted $v_1 \cdots v_n$, i.e., we drop the tensors from the notation. The product on TV is then defined by

$$\mu(u_1 \cdots u_m \otimes v_1 \cdots v_n) = u_1 \cdots u_m v_1 \cdots v_n.$$

The cofree, coassociative coalgebra on V, denoted $\perp V$ in this article, is isomorphic as a graded $\mathbb{Z}$-module to TV. We write $\perp^n V = \bigotimes^n V$, of which a typical basis element is denoted $v_1|\cdots|v_n$. The coproduct on $\perp V$ is then defined in the obvious manner by

$$\begin{aligned}\Delta(v_1|\cdots|v_n) = {} & v_1|\cdots|v_n \otimes 1 + 1 \otimes v_1|\cdots|v_n \\ & + \sum_{i=1}^{n-1} v_1|\cdots|v_i \otimes v_{i+1}|\cdots|v_n.\end{aligned}$$

Let (C,d) be a simply-connected (co)chain coalgebra with reduced coproduct $\overline{\Delta}$. The *cobar construction* on (C,d), denoted $\Omega(C,d)$, is the (co)chain algebra $(Ts^{-1}(C_+), d_\Omega)$, where $d_\Omega = -s^{-1}ds + (s^{-1} \otimes s^{-1})\overline{\Delta}s$ on generators.

Let (A,d) be a connected chain algebra or a simply-connected cochain algebra over R, and let $\bar{A}$ be the component of A of positive degree. The *bar construction* on (A,d), denoted $\mathcal{B}(A,d)$, is a differential graded coalgebra $(\perp(s\bar{A}), D_\mathcal{B})$. Let $(D_\mathcal{B})_1$ denote the linear part of the differential, i.e., $(D_\mathcal{B})_1 = \pi D_\mathcal{B}$, where $\pi : \perp V \to V$ is the natural projection. The linear part of $D_\mathcal{B}$ specifies the entire differential and is given by

$$(D_\mathcal{B})_1(sa_1|\cdots|sa_n) = \begin{cases} -s(da_1) & \text{if} \quad n=1 \\ (-1)^{a_1+1}s(a_1 \cdot a_2) & \text{if} \quad n=2 \\ 0 & \text{if} \quad n>2. \end{cases}$$

Definition 1.1.1. Let $f, g : (A,d) \to (B,d)$ be two maps of chain (respectively, cochain) algebras. An *(f,g)-derivation homotopy* is a linear map $\varphi : A \to B$ of degree $+1$ (respectively, -1) such that $d\varphi + \varphi d = f - g$ and $\varphi\mu = \mu(\varphi \otimes g + f \otimes \varphi)$, where μ denotes the multiplication on A and B.

If f and g are maps of (co)chain coalgebras, there is an obvious dual definition of an *(f,g)-coderivation homotopy*.

We often apply Einstein's summation convention in these notes. When an index appears as both a subscript and a superscript in an expression, it is understood that we sum over that index. For example, given an element c of a coalgebra (C,Δ), the notation $\Delta(c) = c_i \otimes c^i$ means $\Delta(c) = \sum_{i \in I} c_i \otimes c^i$.

Another convention used consistently throughout these notes is the Koszul sign convention for commuting elements of a graded module or for commuting a morphism of graded modules past an element of the source module. For example, if V and W are graded algebras and $v \otimes w, v' \otimes w' \in V \otimes W$, then

$$(v \otimes w) \cdot (v' \otimes w') = (-1)^{|w| \cdot |v'|} vv' \otimes ww'.$$

Futhermore, if $f : V \to V'$ and $g : W \to W'$ are morphisms of graded modules, then for all $v \otimes w \in V \otimes W$,

$$(f \otimes g)(v \otimes w) = (-1)^{|g|\cdot|v|} f(v) \otimes g(w).$$

The source of the Koszul sign convention is the definition of the twisting isomorphism

$$\tau : V \otimes W \to W \otimes V : v \otimes w \mapsto (-1)^{|v|\cdot|w|} w \otimes v.$$

We assume throughout these notes that the reader is familiar with the elements of the theory of simplicial sets and of model categories. We recall here only a few very basic definitions, essentially to fix notation and terminology, and refer the reader to, e.g., [May] and [GJ] for simplicial theory and to [Ho], [DS] and [H] for model category theory.

Definition 1.1.2. Let K be a simplicial set, and let $\mathcal{F}_{ab}$ denote the free abelian group functor. For all $n > 0$, let $DK_n = \cup_{i=0}^{n-1} s_i(K_{n-1})$, the set of degenerate n-simplices of K. The *normalized chain complex* on K, denoted $C_*(K)$, is given by

$$C_n(K) = \mathcal{F}_{ab}(K_n)/\mathcal{F}_{ab}(DK_n).$$

Given a map of simplicial sets $f : K \to L$, the induced map of normalized chain complexes is denoted C_*f.

Recall that $H_*(C_*(K)) \cong H_*(|K|)$ as graded coalgebras, where $|K|$ denotes the geometric realization of K.

Definition 1.1.3. Let K be a reduced simplicial set, and let $\mathcal{F}$ denote the free group functor. The *loop group* GK on K is the simplicial group such that $(GK)_n = \mathcal{F}(K_{n+1} \smallsetminus Im s_0)$, with faces and degeneracies specified by

$$\begin{aligned} \partial_0 \bar{x} &= (\,\overline{\partial_0 x}\,)^{-1} \overline{\partial_1 x} \\ \partial_i \bar{x} &= \overline{\partial_{i+1} x} \quad \text{for all } i > 0 \\ s_i \bar{x} &= \overline{s_{i+1} x} \quad \text{for all } i \geq 0 \end{aligned}$$

where $\bar{x}$ denotes the class in $(GK)_n$ of $x \in K_{n+1}$.

Recall that $H_*(GK) \cong H_*(\Omega|K|)$ as graded Hopf algebras.

In any model category we use the notation $\rightarrowtail$ for cofibrations, $\twoheadrightarrow$ for fibrations and $\xrightarrow{\sim}$ for weak equivalences.

1.2 The canonical, enriched Adams-Hilton model

We recall in this section the construction given in [HPST] of the canonical, enriched Adams-Hilton model of a 1-reduced simplicial set K, upon which our free loop space model construction is based. We begin by reminding the reader of the theories that are essential to this construction. We first sketch briefly the classical and crucial theory of twisting cochains, which goes back to work of E. Brown [Br]. We then outline the theory of strongly homotopy coalgebra maps. We conclude this section by presenting the canonical Adams-Hilton model.

1.2.1 Twisting cochains

Definition 1.2.1. Let (C, d) be a chain coalgebra with coproduct Δ, and let (A, d) be a chain algebra with product μ. A *twisting cochain* from (C, d) to (A, d) is a degree -1 map $t : C \to A$ of graded modules such that

$$dt + td = \mu(t \otimes t)\Delta.$$

The definition of a twisting cochain $t : C \to A$ is formulated precisely so that the following two constructions work smoothly. First, let $(A, d) \otimes_t (C, d) = (A \otimes C, D_t)$, where $D_t = d \otimes 1_C + 1_A \otimes d - (\mu \otimes 1_C)(1_A \otimes t \otimes 1_C)(1_A \otimes \Delta)$. It is easy to see that $D_t^2 = 0$, so that $(A, d) \otimes_t (C, d)$ is a chain complex, which extends (A, d), i.e., of which (A, d) is subcomplex. Second, if C is connected, let $\tilde{t} : Ts^{-1}C_+ \to A$ be the algebra map given by $\tilde{t}(s^{-1}c) = t(c)$. Then $\tilde{t}$ is in fact a chain algebra map $\tilde{t} : \Omega(C, d) \to (A, d)$. It is equally clear that any algebra map $\theta : \Omega(C, d) \to (A, d)$ gives rise to a twisting cochain via the composition

$$C_+ \xrightarrow{s^{-1}} s^{-1}C_+ \hookrightarrow Ts^{-1}C_+ \xrightarrow{\theta} A.$$

Furthermore, the complex $(A, d) \otimes_t (C, d)$ is acyclic if and only if $\tilde{t}$ is a quasi-isomorphism.

The twisting cochain associated to the cobar construction is a fundamental example of this notion. Let (C, d, Δ) be a simply-connected chain coalgebra. Consider the linear map

$$t_\Omega : C \to \Omega C : c \to s^{-1}c.$$

It is an easy exercise to show that t_Ω is a twisting cochain and that $\tilde{t_\Omega} = 1_{\Omega C}$. Thus, in particular, $(\Omega C, d) \otimes_{t_\Omega} (C, d)$ is acyclic; this is the well-known acyclic cobar construction.

1.2.2 Strongly homotopy coalgebra and comodule maps

In [GM] Gugenheim and Munkholm showed that Cotor was natural with respect to a wider class of morphisms than the usual morphisms of chain coalgebras. Given two chain coalgebras (C, d, Δ) and (C', d', Δ'), a *strongly homotopy coalgebra (SHC) map* $f : (C, d, \Delta) \Rightarrow (C', d', \Delta')$ is a chain map $f : (C, d) \to (C', d')$ together with a family of $\mathbb{Z}$-linear maps

$$\mathfrak{F}(f) = \{F_k : C \to (C')^{\otimes k} \mid \deg F_k = k - 1, \ k \geq 1\}$$

satisfying

(1) $F_1 = f$ and

(2) for all $k \geq 2$

$$\begin{aligned} F_k d - \sum_{i+j=k-1} &(-1)^j (1_{C'}^{\otimes i} \otimes d' \otimes 1_{C'}^{\otimes j}) F_k \\ &= \sum_{i+j=k} (-1)^j (F_i \otimes F_j)\Delta + \sum_{i+j=k-2} (-1)^j (1_{C'}^{\otimes i} \otimes \Delta' \otimes 1_{C'}^{\otimes j}) F_{k-1}. \end{aligned}$$

We call $\mathfrak{F}(f)$ an *SHC family* for f.

An SHC map is thus a coalgebra map, up to an infinite family of homotopies. In particular if f is a map of chain coalgebras, then it can be seen as an SHC map, with $F_k = 0$ for all $k > 1$. Furthermore if $f : (C, d, \Delta) \Rightarrow (C', d', \Delta')$ is an SHC map and $g : (C', d', \Delta') \to (C'', d'', \Delta'')$ is a strict coalgebra map, then gf is an SHC map, where $\mathfrak{F}(gf) = \{g^{\otimes k} F_k \mid k \geq 1\}$.

Observe that the existence of $\mathfrak{F}(f)$ is equivalent to the existence of a chain algebra map $\widetilde{\Omega} f : \Omega(C, d) \to \Omega(C', d')$ such that $\widetilde{\Omega} f(s^{-1}c) - s^{-1} f(c) \in T^{\geq 2} s^{-1} C'_+$. Given $\mathfrak{F}(f)$, we can define $\widetilde{\Omega} f$ by setting

$$\widetilde{\Omega} f(s^{-1}c) = \sum_{k \geq 1} (s^{-1})^{\otimes k} F_k(c)$$

and extending to a map of algebras. Condition (2) above then implies that $\widetilde{\Omega} f$ is a differential map as well.

Note that if f is a strict coalgebra map, seen as an SHC map with trivial SHC family, then $\widetilde{\Omega} f = \Omega f$. More generally, if $f : (C, d, \Delta) \Rightarrow (C', d', \Delta')$ is an SHC map and $g : (C', d', \Delta') \to (C'', d'', \Delta'')$ is a strict coalgebra map, seen as an SHC map with trivial SHC family, then there is an SHC family for gf such that $\widetilde{\Omega}(gf) = \Omega g \circ \widetilde{\Omega} f$.

Similarly, given $\widetilde{\Omega} f$, we can define F_k via the composition

$$C_+ \xrightarrow{s^{-1}} s^{-1} C_+ \hookrightarrow T s^{-1} C_+ \xrightarrow{\widetilde{\Omega} f} T s^{-1} C'_+ \xrightarrow{proj} (s^{-1} C'_+)^{\otimes k} \xrightarrow{s^{\otimes k}} (C')^{\otimes k}.$$

Gugenheim and Munkholm proved in [GM] that the usual simplicial Alexander-Whitney map

$$f_{K,L} : C_*(K \times L) \longrightarrow C_*(K) \otimes C_*(L)$$

defined by $f_{K,L}(x, y) = \sum_{i=0}^n \partial_{i+1} \cdots \partial_n x \otimes \partial_0^i y$ is naturally an SHC map.

1.2.3 The canonical Adams-Hilton model

For every pair of simply-connected chain coalgebras (C, d) and (C', d'), Milgram proved that there is a quasi-isomorphism of chain algebras

$$\rho : \Omega\big((C, d) \otimes (C', d')\big) \to \Omega(C, d) \otimes \Omega(C', d') \tag{1.1}$$

specified by $\rho\big(s^{-1}(x \otimes 1)\big) = s^{-1}x$, $\rho\big(s^{-1}(1 \otimes y)\big) = s^{-1}y$ and $\rho\big(s^{-1}(x \otimes y)\big) = 0$ for all $x \in C_+$ and $y \in C'_+$ [Mi].

In [S] Szczarba gave an explicit formula for a natural transformation of functors from simplicial sets to chain algebras

$$\theta : \Omega C_*(-) \to C_*(G(-))$$

such that $\theta_K : \Omega C_*(K) \to C_*(GK)$ is a quasi-isomorphism of chain algebras for every 1-reduced simplicial set K. Since $C_*(GK)$ is in fact a chain Hopf algebra, it is reasonable to ask whether $\Omega_* C(K)$ can be endowed with a coproduct with respect to which θ_K is a quasi-isomorphism of chain Hopf algebras.

Let $\psi : \Omega C_*(-) \to \Omega C_*(-) \otimes \Omega C_*(-)$ denote the natural transformation given for each 1-reduced simplicial set K by the composition

$$\Omega C_*(K) \xrightarrow{\Omega(\Delta_K)_\sharp} \Omega C_*(K \times K) \xrightarrow{\widetilde{\Omega} f_{K,K}} \Omega\big(C_*(K) \otimes C_*(K)\big) \xrightarrow{\rho} \Omega C_*(K) \otimes \Omega C_*(K).$$

The coproduct $\psi_K : \Omega C_*(K) \to \Omega C_*(K) \otimes \Omega C_*(K)$ is called the *Alexander-Whitney (A-W) cobar diagonal.* In [HPST] Hess, Parent, Scott and Tonks proved that for all 1-reduced K, the Alexander-Whitney cobar diagonal is strictly coassociative and cocommutative up to derivation homotopy, which we call Θ. They established furthermore that Szczarba's equivalence θ_K is an SHC map with respect to ψ_K and the usual coproduct on $C_*(GK)$.

In [B] Baues provided a purely combinatorial definition of strictly coassociative coproduct and of a derivation homotopy for cocommutativity on $\Omega C_*(K)$ for any 1-reduced simplicial set K, but without giving a map from $\Omega C_*(K)$ to $C_*(GK)$. In [HPST] it is shown that the Alexander-Whitney cobar diagonal is the same as Baues's coproduct, which implies that

$$\operatorname{Im} \overline{\psi}_K \subseteq T^{\geq 1} s^{-1} C_+ K \otimes s^{-1} C_+ K,$$

where $\overline{\psi}_K$ is the reduced coproduct.

Henceforth we refer to $\theta_K : \Omega C_*(K) \to C_*(GK)$ as the *canonical Adams-Hilton model* and to $\psi_K : \Omega C_*(K) \to \Omega C_*(K) \otimes \Omega C_*(K)$ as its *canonical enrichment.*

1.3 Noncommutative algebraic models of fiber squares

We review in this section the bare essentials of noncommutative modeling of fiber squares, as developed in [DH1]. Note that this theoretical framework is highly analogous to the theory of KS-extensions in rational homotopy theory. See [DH1]and [FHT] for more details.

We first define the classes of morphisms with which we work throughout the remainder of this article.

Definition 1.3.1. Let (B, d) and (C, d) be bimodules over an associative cochain algebra (A, d). A cochain map $f : (B, d) \longrightarrow (C, d)$ is a *quasi-bimodule map* if H^*f is a map of $\mathrm{H}^*(A, d)$-bimodules. If (A, d) and (A', d') are associative cochain algebras, then a cochain map $f : (A, d) \longrightarrow (A', d')$ is a *quasi-algebra map* if H^*f is a map of algebras.

Noncommutative cochain algebra models of topological spaces are defined in terms of quasi-algebra maps.

Definition 1.3.2. Let X be a topological space. An (integral) *noncommutative model* of X consists of an associative cochain algebra over $\mathbb{Z}$, (A, d), together with a quasi-algebra quasi-isomorphism

$$\alpha : (A, d) \xrightarrow{\simeq} C^*(X) ,$$

where α is called a *model morphism.*

Of course, we must also define what it means to model a continuous map, if we wish to model pull-backs of fibrations.

Definition 1.3.3. Let $f : Y \to X$ be a continuous map. A *noncommutative model* of f consists of a commuting diagram

$$\begin{array}{ccc} (A,d) & \xrightarrow{\varphi} & (B,d) \\ \simeq \downarrow \alpha & & \simeq \downarrow \beta \\ C^*X & \xrightarrow{C^*f} & C^*Y \end{array}$$

in which α and β are model morphisms, and φ is a quasi-algebra map.

Remark 1.3.4. (1) In most applications of noncommutative models, α and φ are strict morphisms of algebras, while it is usually impossible for β to be a strict morphism of algebras.

(2) It is often difficult to use a noncommutative model for constructions or calculations, unless A is a free algebra. It is easy to see, however, that every space X possesses such a noncommutative model.

There is a special class of strict algebra maps, known as *twisted algebra extensions*, that are used for modeling topological fibrations. Roughly speaking, a twisted algebra extension of one algebra by another is a tensor product of the two algebras in which both the differential and multiplication are perturbed from the usual tensor-product differential and multiplication.

Definition 1.3.5. Let (A, d) and (B, d) be a cochain algebra and a cochain complex over $\mathbb{Z}$, respectively. A *twisted bimodule extension* of (A, d) by (B, d) is an (A, d)-bimodule (C, D) such that

(1) $C \cong A \otimes B$ as graded modules;

(2) the right action of A on C is free, i.e., $(a \otimes b) \cdot a' = (-1)^{a'b} aa' \otimes b$;

(3) the left action of A on C commutes with the right action, i.e.,

$$(a \cdot c) \cdot a' = a \cdot (c \cdot a')$$

for all $a, a' \in A$ and $c \in C$, and satisfies

$$a \cdot (1 \otimes b) - a \otimes b \in A^+ \otimes B^{<b}$$

for all a in A and b in B; and

(4) the inclusion map $(A, d) \to (C, D)$ and the projection map $(C, D) \to (B, d)$ are both (A, d)-bimodule maps, where (B, d) is considered with the trivial (A, d)-bimodule structure, so that, in particular,

$$D(1 \otimes b) - 1 \otimes db \in A^+ \otimes B$$

for all b, b' in B.

If (B, d) is a cochain algebra, then a twisted bimodule extension (C, D) of (A, d) by (B, d) is a *twisted algebra extension* if the bimodule structure of (C, D) extends to a full algebra structure such that the inclusion and projection maps above are maps of cochain algebras. In particular,

$$(1 \otimes b)(1 \otimes b') - 1 \otimes bb' \in A^+ \otimes B$$

for all b, b' in B.

Notation 1.3.6. We write $(A, d)\widetilde{\otimes}(B, d)$ to denote a twisted bimodule extension of (A, d) by (B, d) and $(A, d) \odot (B, d)$ to denote a twisted algebra extension.

The proposition below, which is the noncommutative analogue of a well-known result concerning KS-extensions, states that twisted algebra extensions have the left lifting property with respect to surjective quasi-algebra quasi-isomorphisms. Since it is natural to think of surjective quasi-algebra morphisms as fibrations of cochain algebras, Proposition 1.3.7 implies that we can think of twisted extensions as cofibrations. In other words, twisted algebra extensions are plausible models of topological fibrations, since the cochain functor is contravariant.

Proposition 1.3.7. *Let $\iota : (A,d) \to (A,d) \odot (B,d)$ be a twisted algebra extension. Given a commuting diagram*

$$\begin{array}{ccc} (A,d) & \xrightarrow{f} & (C,d) \\ \downarrow{\scriptstyle \iota} & & \simeq \downarrow{\scriptstyle p} \\ (A,d)\odot(B,d) & \xrightarrow{g} & (E,d) \end{array}$$

in which f is a right (A,d)-module map, p is a surjective quasi-algebra quasi-isomorphism, and g is a quasi-algebra map, there exists a quasi-algebra map,

$$h : (A,d) \odot (B,d) \longrightarrow (C,d)$$

which is a right (A,d)-module map, as well as a lift of g through p and an extension of f, i.e., $ph = g$ and $h\iota = f$.

This proposition is a simplified version of a result that first appeared in [DH1], but that we do not need in its full generality here.

Proof. Since $(A,d) \odot (B,d)$ is semifree as a right (A,d)-module, the lift h exists as a map of right (A,d)-modules. In cohomology $\mathrm{H}^* p\, \mathrm{H}^* h = \mathrm{H}^* g$, which implies that $\mathrm{H}^* h = (\mathrm{H}^* p)^{-1}\, \mathrm{H}^* g$, since $\mathrm{H}^* p$ is an isomorphism. Hence $\mathrm{H}^* h$ is an algebra map, as it is a composition of algebra maps. □

Let us see how to model pull-backs of fibrations in this context. Consider a pull-back square of simply-connected spaces

$$\begin{array}{ccc} E \times_B X & \xrightarrow{\bar f} & E \\ \downarrow{\scriptstyle \bar q} & & \downarrow{\scriptstyle q} \\ X & \xrightarrow{f} & B \end{array}$$

in which q is a fibration and f an arbitrary continuous map. Suppose that

$$\begin{array}{ccc} (A,d) & \xrightarrow{\varphi} & (\bar A,\bar d) \\ \simeq\downarrow{\scriptstyle \alpha} & & \simeq\downarrow{\scriptstyle \gamma} \\ C^*B & \xrightarrow{C^*f} & C^*X \end{array} \tag{1.2}$$

and

$$\begin{array}{ccc} (A,d) & \xrightarrow{\iota} & (A,d)\odot(C,e) \\ \simeq\downarrow{\scriptstyle \alpha} & & \simeq\downarrow{\scriptstyle \beta} \\ C^*B & \xrightarrow{C^*q} & C^*E \end{array} \tag{1.3}$$

are noncommutative models of f and q, where ι is a twisted algebra extension of (A, d). We assume that α and γ are algebra maps, while β may be only a quasi-algebra map.

The following theorem provides the theoretical underpinnings for noncommutative modeling of fiber squares. It states that under certain reasonable conditions, there exists a sort of push-out of ι and φ that is a model of the pull-back $E \underset{B}{\times} X$.

Theorem 1.3.8. [DH1] *Given a commuting diagram over a field* $\Bbbk$, *with squares as in diagrams* 1.2 *and* 1.3,

$$\begin{array}{ccccc} (A,d)\odot(C,e) & \xleftarrow{\iota} & (A,d) & \xrightarrow{\varphi} & (\bar{A},\bar{d}) \\ \simeq\downarrow\beta & & \simeq\downarrow\alpha & & \simeq\downarrow\gamma \\ C^*E & \xleftarrow{C^*q} & C^*B & \xrightarrow{C^*f} & C^*X \end{array}$$

in which $\bar{A}$ *is a free algebra and* φ *admits a cochain algebra section* σ, *there exist a twisted algebra extension*

$$\bar{\iota} : (\bar{A}, \bar{d}) \to (\bar{A}, \bar{d}) \odot (C, e)$$

and a noncommutative model over $\Bbbk$

$$\delta : (\bar{A}, \bar{d}) \odot (C, e) \xrightarrow{\simeq} C^*(E \underset{B}{\times} X)$$

such that

(1) $(\bar{a} \otimes 1)(1 \otimes c) = (\varphi \otimes 1)\big((\sigma(\bar{a}) \otimes 1)(1 \otimes c)\big)$ *for all* $\bar{a}$ *in* $\bar{A}$ *and* c *in* C;

(2) *if* D *is the differential on* $A \otimes C$, *then the differential* $\bar{D}$ *on* $\bar{A} \otimes C$ *commutes with the right action of* $\bar{A}$ *and is specified by* $\bar{D}(1 \otimes c) = (\varphi \otimes 1)\big(D(1 \otimes c)\big)$ *for all* c *in* C;

(3) $\delta(\bar{a} \otimes c) = C^*\bar{f} \circ \beta(\sigma(\bar{a}) \otimes c)$ *for all* $\bar{a}$ *in* $\bar{A}$ *and* c *in* C.

When we are not working over a field, as in this chapter, we cannot apply this theorem directly but have to employ more ad hoc methods, in order to obtain a result of this type. In particular, defining the full algebra structure on $(\bar{A}, \bar{d}) \otimes (C, e)$ and then showing that δ is a quasi-algebra map can be delicate.

Related work 1.3.9. As mentioned at the beginning of this subsection, our approach to algebraic modeling of fiber squares is analogous to the KS-extensions of rational homotopy theory, as developed by Sullivan, see [FHT]. The *Adams-Hilton model*, which to any 1-reduced CW-complex X associates a chain algebra $(AH(X), d)$ quasi-isomorphic to the cubical chains on ΩX [AH], is another particularly useful tool for algebraic modeling. The algebra $AH(X)$ is free on generators in one-to-one correspondence with the cells of X, and the differential d encodes the attaching maps.

In [An], Anick showed that the Adams-Hilton model could be endowed with a coproduct ψ, so that it became a *Hopf algebra up to homotopy*. He showed furthermore that if X is a finite r-connected CW complex of dimension at most rp, then there is a commutative cochain algebra $A(X)$ that is quasi-isomorphic to $C^*(X;\mathbb{F}_p)$, the algebra of mod p cochains on X. Using Anick's result, Menichi proved in [Me] that if $i : X \hookrightarrow Y$ is an inclusion of finite r-connected CW complexes of dimension at most rp and F is the homotopy fiber of i, then the mod p cohomology of F is isomorphic as an algebra to $\mathrm{Tor}^{A(Y)}(A(X), \mathbb{F}_p)$.

Other interesting algebraic models include the *SHC-algebras* studied in [NT] by Ndombol and Thomas and E_∞*-algebras*, shown by Mandell to serve as models for p-complete homotopy theory [Man]. In particular, Mandell proved that the cochain functor $C^*(-;\overline{\mathbb{F}_p})$ embeds the category of nilpotent p-complete spaces onto a full subcategory of E_∞-algebras. He also characterized those E_∞-algebras that are weakly equivalent to the cochains on a p-complete space.

Chapter 2

Free loop spaces

Consider the free loop fiber square for a simply-connected CW-complex of finite type, X.

$$\begin{array}{ccc} \mathcal{L}X & \xrightarrow{j} & X^I \\ \big\downarrow{\scriptstyle e} & & \big\downarrow{\scriptstyle (\mathrm{ev}_0,\mathrm{ev}_1)} \\ X & \xrightarrow{\Delta} & X \times X \end{array}$$

Here ev_t is defined by $\mathrm{ev}_t(\ell) = \ell(t)$, and Δ is the diagonal.

Our goal in this chapter is to construct canonically an associative cochain algebra $fls^*(X)$ together with a quasi-isomorphism $fls^*(X) \xrightarrow{\simeq} C^*\mathcal{L}X$ that induces an isomorphism of algebras in cohomology.

We construct the noncommutative model of $\mathcal{L}X$ over $\mathbb{Z}$ in three steps. First we find a model of Δ, a relatively easy exercise. The second step, in which we define a twisted extension of cochain algebras that is a model of the topological fibration $(\mathrm{ev}_0, \mathrm{ev}_1)$, requires considerably more work. Once we have obtained the models of Δ and $(\mathrm{ev}_0, \mathrm{ev}_1)$, we show that they can be "twisted together," leading to a model for $\mathcal{L}X$.

Since it is much easier to obtain precise, natural algebraic models for simplicial sets than for topological spaces, we begin this chapter by constructing a useful, canonical simplicial model of the free loop space. We then apply the general theory of section 1.3 to building the desired algebraic free loop space model.

2.1 A simplicial model for the free loop space

2.1.1 The general model

Let $\mathcal{S}_\bullet : \mathcal{T}op \longrightarrow s\mathcal{S}et$ denote the singular simplicial set functor, which is a right adjoint to the geometric realization functor, with which it forms a Quillen equivalence. Let $\eta : Id \longrightarrow \mathcal{S}_\bullet|\cdot|$ denote the unit of the adjunction. Recall that η_L is always a weak equivalence [May].

Let X be a 1-connected space, and let K be a 1-reduced Kan complex such that $|K|$, the geometric realization of K, has the homotopy type of X. For example, we could take $K = \mathcal{S}_\bullet(X)$. Let

$$\begin{array}{ccc} K & \xrightarrow{\quad\Delta\quad} & K \times K \\ & {\scriptstyle i,\,\simeq}\searrow \qquad \nearrow{\scriptstyle p} & \\ & P & \end{array}$$

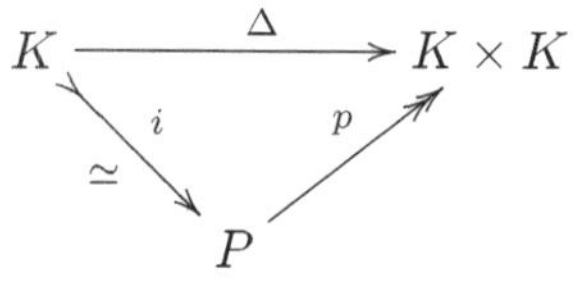

be a path object on K. Set $\mathcal{L}(K,P) := P \underset{K\times K}{\times} K$. Let $\bar{p} : \mathcal{L}(K,P) \longrightarrow K$ and $\overline{\Delta} : \mathcal{L}(K,P) \longrightarrow P$ denote the canonical maps, i.e.,

$$\begin{array}{ccc} \mathcal{L}(K,P) & \xrightarrow{\ \overline{\Delta}\ } & P \\ \downarrow{\scriptstyle \bar{p}} & & \downarrow{\scriptstyle p} \\ K & \xrightarrow{\ \Delta\ } & K \times K \end{array}$$

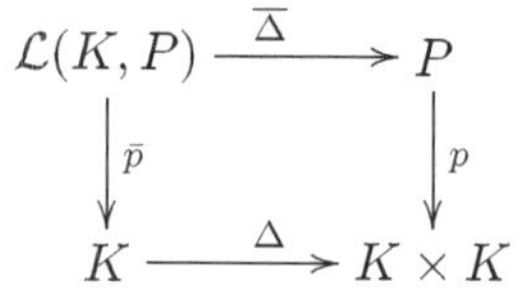

is the pullback diagram.

Proposition 2.1.1. *There is a weak equivalence* $\mathcal{L}(K,P) \longrightarrow \mathcal{S}_\bullet(\mathcal{L}X)$.

Proof. Since $\mathcal{S}_\bullet$ is a right adjoint, it preserves limits. It also preserves fibrations, as the right half of a Quillen equivalence. There is therefore a pullback diagram

$$\begin{array}{ccc} \mathcal{S}_\bullet(\mathcal{L}|K|) & \xrightarrow{\quad \mathcal{S}_\bullet(j) \quad} & \mathcal{S}_\bullet(|K|^I) \\ \downarrow{\scriptstyle \mathcal{S}_\bullet(e)} & & \downarrow{\scriptstyle (\mathcal{S}_\bullet(\mathrm{ev}_0),\mathcal{S}_\bullet(\mathrm{ev}_1))} \\ \mathcal{S}_\bullet(|K|) & \xrightarrow{\quad \mathcal{S}_\bullet(\Delta) \quad} & \mathcal{S}_\bullet(|K|) \times \mathcal{S}_\bullet(|K|) \end{array}$$

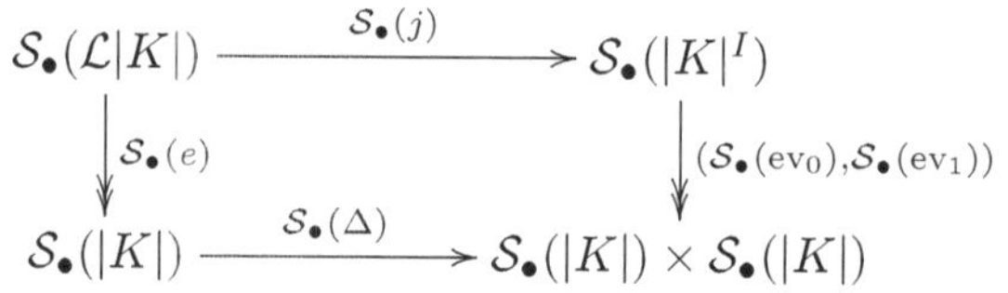

since $\mathcal{S}_\bullet(|K| \times |K|) \cong \mathcal{S}_\bullet(|K|) \times \mathcal{S}_\bullet(|K|)$. Consider the following diagram, which commutes by naturality of η,

$$\begin{array}{ccccc} K & \underset{\simeq}{\xrightarrow{\eta_K}} & \mathcal{S}_\bullet(|K|) & \underset{\simeq}{\xrightarrow{\quad \mathcal{S}_\bullet(s) \quad}} & \mathcal{S}_\bullet(|K|^I) \\ {\scriptstyle \simeq}\downarrow{\scriptstyle i} & \searrow{\scriptstyle \Delta} & & \searrow{\scriptstyle \mathcal{S}_\bullet(\Delta)} & \downarrow{\scriptstyle (\mathcal{S}_\bullet(\mathrm{ev}_0),\mathcal{S}_\bullet(\mathrm{ev}_1))} \\ P & \xrightarrow{\ p\ } & K \times K & \underset{\simeq}{\xrightarrow{\quad \eta_{K^2} \quad}} & \mathcal{S}_\bullet(|K|) \times \mathcal{S}_\bullet(|K|) \end{array}$$

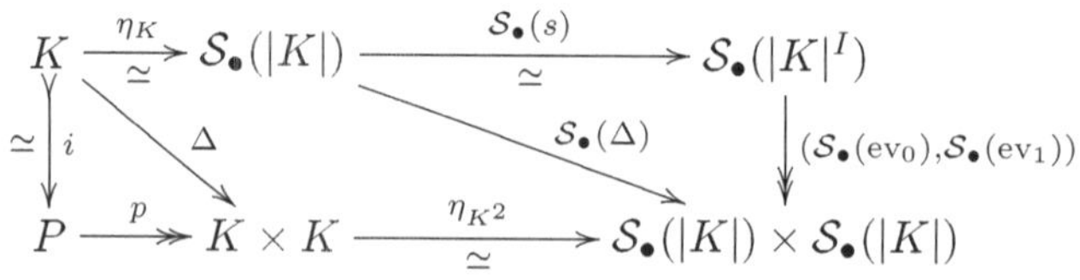

where $s : |K| \longrightarrow |K|^I$ is the usual section, sending an element x of $|K|$ to the constant path at x. Since i is an acyclic cofibration and $(\mathcal{S}_\bullet(\mathrm{ev}_0), \mathcal{S}_\bullet(\mathrm{ev}_1))$ is a fibration, there is a simplicial map $\bar{\eta} : P \longrightarrow \mathcal{S}_\bullet(|K|^I)$ such that

$$\begin{array}{ccc} K & \xrightarrow[\simeq]{\mathcal{S}_\bullet(s)\eta_K} & \mathcal{S}_\bullet(|K|^I) \\ {\scriptstyle \simeq}\big\downarrow{\scriptstyle i} & \nearrow^{\bar{\eta}} & \big\downarrow{\scriptstyle (\mathcal{S}_\bullet(\mathrm{ev}_0),\mathcal{S}_\bullet(\mathrm{ev}_1))} \\ P & \xrightarrow{\eta_{K^2}\circ p} & \mathcal{S}_\bullet(|K|) \times \mathcal{S}_\bullet(|K|) \end{array}$$

commutes. Note that by "2-out-of-3" $\bar{\eta}$ is a weak equivalence.

We have therefore a commutative diagram

$$\begin{array}{ccccc} \mathcal{L}(K,P) & & \xrightarrow{\bar{\eta}\overline{\Delta}} & & \\ & \searrow^{\eta_K\bar{p}} & & & \\ & \mathcal{S}_\bullet(\mathcal{L}|K|) & \xrightarrow{\mathcal{S}_\bullet(j)} & \mathcal{S}_\bullet(|K|^I) & \\ & \big\downarrow{\scriptstyle \mathcal{S}_\bullet(e)} & & \big\downarrow{\scriptstyle (\mathcal{S}_\bullet(\mathrm{ev}_0),\mathcal{S}_\bullet(\mathrm{ev}_1))} & \\ & \mathcal{S}_\bullet(|K|) & \xrightarrow{\mathcal{S}_\bullet(\Delta)} & \mathcal{S}_\bullet(|K|) \times \mathcal{S}_\bullet(|K|) & \end{array}$$

which implies that there is a simplicial map

$$\mathcal{L}(K,P) \xrightarrow{\hat{\eta}} \mathcal{S}_\bullet(\mathcal{L}|K|)$$

such that $\mathcal{S}_\bullet(j)\hat{\eta} = \bar{\eta}\overline{\Delta}$ and $\mathcal{S}_\bullet(e)\hat{\eta} = \eta_K\bar{p}$. By the "Cogluing Lemma" (cf. [GJ, §II.8]), it is clear that $\hat{\eta}$ is a weak equivalence. □

For explicit computations to be possible, it is important to be able to build a simplicial model for $\mathcal{L}X$ from a simplicial set that isn't necessarily Kan. Suppose therefore that K is any 1-reduced simplicial set such that $|K| \simeq X$. Let

$$K \underset{\simeq}{\overset{j}{\rightarrowtail}} K' \twoheadrightarrow \{*\}$$

be a fibrant replacement of K.

Given path objects on K and on K',

$$K \underset{\simeq}{\overset{i}{\rightarrowtail}} P \overset{p}{\twoheadrightarrow} K \times K$$

and

$$K' \underset{\simeq}{\overset{i'}{\rightarrowtail}} P' \overset{p'}{\twoheadrightarrow} K' \times K',$$

consider the commuting square

$$\begin{array}{ccccc} K & \xrightarrow[\simeq]{j} & K' & \xrightarrow[\simeq]{i'} & P' \\ {\scriptstyle \simeq}\big\downarrow{\scriptstyle i} & & & & \big\downarrow{\scriptstyle p'} \\ P & \xrightarrow{p} & K \times K & \xrightarrow[\simeq]{j\times j} & K' \times K' \end{array}$$

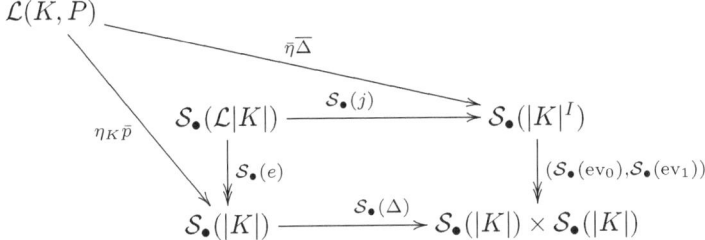

in which $j \times j$ is a weak equivalence. Since i is an acyclic cofibration and p' is a fibration, there is a simplicial map $\tilde{\jmath} : P \longrightarrow P'$ such that

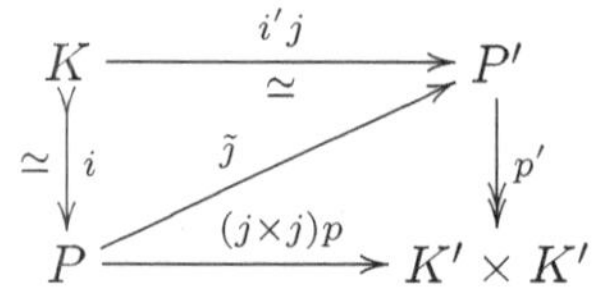

commutes. By "2-out-of-3", $\tilde{\jmath}$ is also weak equivalence.

If, as above, $\mathcal{L}(K,P) := P \underset{K\times K}{\times} K$ and $\mathcal{L}(K',P') := P' \underset{K'\times K'}{\times} K'$, then by the universal property of pull-backs, there is a simplicial map $\tilde{\jmath}' : \mathcal{L}(K,P) \longrightarrow \mathcal{L}(K',P')$ such that the following cube commutes.

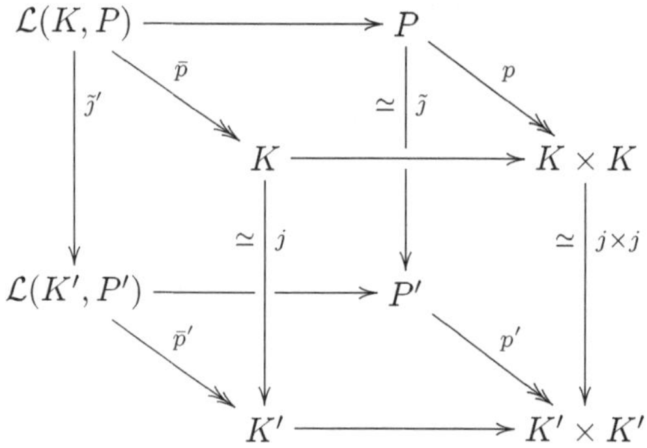

As in the proof of Proposition 2.1.1, the "Cogluing Lemma" implies that $\tilde{\jmath}'$ is a weak equivalence. In particular, $\mathrm{H}^*(\mathcal{L}(K,P)) \cong \mathrm{H}^*(\mathcal{L}(K',P'))$, as algebras. We have thus established the following result.

Theorem 2.1.2. *Let X be a 1-connected space, and let $\mathcal{L}X$ be the free loop space on X. Let K be any 1-reduced simplicial set such that $|K|$ has the homotopy type of X. Let $K \overset{i}{\underset{\simeq}{\rightarrowtail}} P \overset{p}{\twoheadrightarrow} K \times K$ be a path object on K, and set $\mathcal{L}(K,P) := P \underset{K\times K}{\times} K$. Then there is simplicial weak equivalence*

$$\mathcal{L}(K,P) \xrightarrow{\simeq} \mathcal{S}_\bullet(\mathcal{L}X)\ .$$

In particular, $\mathrm{H}^*(\mathcal{L}(K,P)) \cong \mathrm{H}^*(\mathcal{L}X)$ as algebras.

2.1.2 Choosing the free loop model functorially

Let K be a 1-reduced simplicial set. Let GK denote the Kan loop group on K, as defined in the preface. Let PK denote the twisted cartesian product with structure group $GK \times GK$

$$PK := GK \underset{\tau}{\times} (K \times K),$$

where $\tau : K \times K \longrightarrow GK \times GK : (x, y) \longmapsto (\bar{x}, \bar{y})$ and $GK \times GK$ acts on GK by $(v, w) \cdot u := vuw^{-1}$. It is easy to verify that τ satisfies the conditions of a twisted cartesian product. Observe that, in particular,

$$\partial_0(u, (x, y)) = (\bar{x} \cdot \partial_0 u \cdot \bar{y}^{-1}, (\partial_0 x, \partial_0 y)).$$

Proposition 2.1.3. *Let* $p : PK \longrightarrow K \times K$ *denote the projection map, and let*

$$i : K \longrightarrow PK : x \longmapsto (e, x, x).$$

Then $K \xrightarrow{i} PK \xrightarrow{p} K \times K$ *is a path object on* K.

Proof. Since PK is a twisted cartesian product and GK is a Kan complex, p is a Kan fibration. Furthermore, i is a simplicial cofibration, since it is obviously injective. We need therefore only to show that i is a weak equivalence.

If K is a Kan complex, then there is a long exact sequence of homotopy groups

$$\cdots \longrightarrow \pi_{n+1}(K \times K) \xrightarrow{\delta_\tau} \pi_n(GK) \longrightarrow \pi_n(PK) \longrightarrow \pi_n(K \times K) \longrightarrow \cdots,$$

where δ_τ is the connecting homomorphism, defined by $\delta_\tau([x, y]) = [\bar{x} \cdot \bar{y}^{-1}]$ (cf., [GJ, §I.7]). Comparing this long exact sequence with that obtained from the universal acyclic twisted cartesian product $GK \underset{\eta}{\times} K = EK$

$$\cdots \longrightarrow \pi_{n+1}(EK) \longrightarrow \pi_{n+1}(K) \xrightarrow{\delta_\eta} \pi_n(GK) \longrightarrow \pi_n(EK) \longrightarrow \cdots,$$

we obtain that δ_η is an isomorphism given by $\delta_\eta([x]) = [\bar{x}]$, and therefore that δ_τ is surjective. Furthermore, $\ker \delta_\tau = \operatorname{Im} \pi_{n+1}\Delta$, since

$$[x, y] \in \ker \delta_\tau \Longleftrightarrow [\bar{x} \cdot \bar{y}^{-1}] = [e] \Longleftrightarrow [\bar{x}] = [\bar{y}] \Longleftrightarrow [x] = [y] \Longleftrightarrow [x, y] = [x, x]$$

where the second equivalence comes from multiplying both sides of the equation by $[\bar{y}]$, using the multiplication induced by that on GK, while the third equivalence is due to the fact that δ_η is an isomorphism.

The first long exact sequence therefore breaks up into short exact sequences

$$0 \longrightarrow \pi_{n+1}(PK) \xrightarrow{\pi_{n+1}p} \pi_{n+1}(K \times K) \xrightarrow{\delta_\tau} \pi_n(GK) \longrightarrow 0,$$

which are split, since δ_τ has an obvious section σ defined by $\sigma([\bar{x}]) = [x, *]$, where $*$ is (an iterated degeneracy of) the basepoint of K. Consequently $\pi_{n+1}p$ has a left inverse ρ, and so

$$\pi_{n+1}i = \rho\pi_{n+1}p\pi_{n+1}i = \rho\pi_{n+1}\Delta.$$

Since $\operatorname{Im}\pi_{n+1}\Delta = \ker\delta_\tau = \operatorname{Im}\pi_{n+1}p$ and the restriction of ρ to $\operatorname{Im}\pi_{n+1}p$ is surjective, $\pi_{n+1}i$ is surjective as well. Furthermore $\pi_{n+1}i$ is necessarily injective, as $\pi_{n+1}pr_1\pi_{n+1}p$ is a left inverse to $\pi_{n+1}i$, where $pr_1 : K \times K \longrightarrow K$ is the projection map onto the first factor. Thus $\pi_{n+1}i$ is an isomorphism for all n, i.e., i is a weak equivalence.

If K is not a Kan complex, consider a fibrant replacement $K \underset{\simeq}{\overset{j}{\rightarrowtail}} K' \twoheadrightarrow \{*\}$, which, by the naturality of the P-construction, induces a map of twisted cartesian products

$$\begin{array}{ccccc}
GK & \longrightarrow & PK & \overset{p}{\twoheadrightarrow} & K \\
\simeq \big\downarrow Gj & & \big\downarrow Pj & & \simeq \big\downarrow j \\
GK' & \longrightarrow & PK' & \twoheadrightarrow & K'
\end{array}$$

in which Gj is an acyclic cofibration, since G is the left member of a Quillen equivalence. A 5-Lemma argument applied to the long exact sequences in homotopy of the realizations of these fibrations then shows that Pj is a weak equivalence. Again by naturality, the square

$$\begin{array}{ccc}
K & \overset{i}{\longrightarrow} & PK \\
\simeq \big\downarrow j & & \simeq \big\downarrow Pj \\
K' & \underset{\simeq}{\overset{i'}{\longrightarrow}} & PK'
\end{array}$$

commutes and so, by "2-out-ot-3", i is also a weak equivalence. □

We now use this functorial path object construction to define a functorial, simplicial free loop space model.

Definition 2.1.4. Given a 1-reduced simplicial set K, the *canonical free loop construction* on K is the the simplicial set

$$\mathcal{L}K := \mathcal{L}(K, PK) = PK \underset{K\times K}{\times} K.$$

Observe that $\mathcal{L}K$ is the twisted cartesian product $GK \underset{\bar{\tau}}{\times} K$, with structure group $GK \times GK$, where $GK \times GK$ acts on GK as before and

$$\bar{\tau} : K \longrightarrow GK \times GK : x \longmapsto (\bar{x}, \bar{x})$$

so that

$$\partial_0(w, x) = (\bar{x} \cdot \partial_0 w \cdot \bar{x}^{-1}, \partial_0 x).$$

According to Theorem 2.1.2, there is a weak equivalence $\mathcal{L}K \xrightarrow{\simeq} \mathcal{S}_\bullet(\mathcal{L}|K|)$ and therefore an algebra isomorphism $\mathrm{H}^*(\mathcal{L}K) \cong \mathrm{H}^*(\mathcal{L}|K|)$.

2.2 The multiplicative free loop space model

In this section we apply the methods described in section 1.3 to constructing naturally a noncommutative model for the canonical free loop construction $\mathcal{L}K$ on a 1-reduced simplicial set of finite-type K. By Theorem 2.1.2 we obtain therefore a noncommutative model for the free loop space $\mathcal{L}X$ on a 1-connected space X with the homotopy type of a finite-type CW-complex.

We begin by constructing specific explicit models of the diagonal map

$$\Delta : K \longrightarrow K \times K$$

and of the path fibration

$$p : PK \longrightarrow K \times K,$$

which we then twist together appropriately, in order to obtain a model of the free loop construction.

2.2.1 The diagonal map

Let $\varepsilon : \Omega\mathcal{B} \longrightarrow Id$ denote the counit of the cobar-bar adjunction, which is a quasi-isomorphism for each 1-connected cochain algebra (A, d). Let

$$\gamma = \varepsilon_K : \Omega\mathcal{B}C^*K \xrightarrow{\simeq} C^*K \, .$$

Let α denote the following composition.

$$\begin{array}{ccccc} \Omega(\mathcal{B}C^*K \otimes \mathcal{B}C^*K) & \xrightarrow{\Omega(\rho^\sharp)} & \Omega\mathcal{B}(C^*K \otimes C^*K) & \xrightarrow{\Omega(\widetilde{\Omega} f_{K,K})^\sharp} & \Omega\mathcal{B}C^*(K \times K) \\ & \searrow^{\alpha} & & & \simeq \big\downarrow \varepsilon_{K \times K} \\ & & & & C^*(K \times K) \end{array}$$

Here $\rho^\sharp$ is the dual of the map defined in equation 1.1, while $f_{K,K}$ is again the Alexander-Whitney map, as in section 1.2. Note that we are relying on the fact that K is of finite-type, in writing, e.g., $\mathcal{B}C^*K$ for the dual of ΩC_*K.

Given these definitions, it is an easy exercise, using the naturality of ε, to show that

$$\begin{array}{ccc} \Omega(\mathcal{B}C^*K \otimes \mathcal{B}C^*K) & \xrightarrow{\Omega\psi_K^\sharp} & \Omega\mathcal{B}C^*K \\ \simeq \big\downarrow \alpha & & \simeq \big\downarrow \gamma \\ C^*(K \times K) & \xrightarrow{C^*\Delta} & C^*K \end{array} \qquad (2.1)$$

commutes and is therefore a model of Δ. Observe that if

$$\iota_k : \Omega\mathcal{B}C^*K \longrightarrow \Omega(\mathcal{B}C^*K \otimes \mathcal{B}C^*K)$$

is the inclusion of the kth tensor factor $(k = 1, 2)$, then $C^*pr_k \circ \gamma = \alpha\iota_k$.

2.2.2 The path fibration

Henceforth, we employ the following useful notation.

Notation 2.2.1. For any $a, b \in \mathcal{B}C^*K$, let $a \star b := \psi_K^\sharp(a \otimes b)$. Furthermore, let $\varphi = \Omega\psi_K^\sharp$.

We begin by defining a certain twisted extension of $\Omega(\mathcal{B}C^*K \otimes \mathcal{B}C^*K)$ by $\mathcal{B}C^*K$. Since the former is a model of $K \times K$, while the latter is a model of GK, it is reasonable to expect to be able to construct a model of PK of this form. Once we have the explicit definition of the twisted extension, we show that it is indeed a model of PK.

Definition 2.2.2. of $\Omega(\mathcal{B}C^*K \otimes \mathcal{B}C^*K) \odot \mathcal{B}C^*K$.
Let $\Omega(\mathcal{B}C^*K \otimes \mathcal{B}C^*K) \odot \mathcal{B}C^*K$ be the twisted algebra extension determined by the following conditions.

(1) The left action of $\Omega(\mathcal{B}C^*K \otimes \mathcal{B}C^*K)$ is specified recursively by

$$\begin{aligned}
&\big(s^{-1}(a \otimes b) \otimes 1\big) \cdot (1 \otimes sx_1|\cdots|sx_n)\\
&=s^{-1}(a \otimes b) \otimes sx_1|\cdots|sx_n\\
&\quad+(-1)^{b\cdot\theta_n}\Big(s^{-1}\big(a \otimes (sx_1|\cdots|sx_n \star b)\big) - s^{-1}\big((a \star sx_1|\cdots|sx_n) \otimes b\big)\Big)\\
&\quad+\sum_{j=1}^{n-1}\Big[(-1)^{b\cdot\theta_j} s^{-1}\big((a \star sx_1|\cdots|sx_j) \otimes b\big) \cdot (1 \otimes sx_{j+1}|\cdots|sx_n)\\
&\qquad+(-1)^{(b+\theta_j)(\theta_n-\theta_j)} s^{-1}\big(a \otimes (sx_{j+1}|\cdots|sx_n) \star b\big) \otimes sx_1|\cdots|sx_j\Big]
\end{aligned}$$

where $\theta_j = j + \sum_{i\leq j}|x_i|$.

(2) The restriction of the differential D of the twisted extension to $1 \otimes \mathcal{B}C^*K$ is specified recursively by

$$\begin{aligned}
D(1\otimes sx_1|\cdots|sx_n) &= 1 \otimes d_{\mathcal{B}}(sx_1|\cdots|sx_n)\\
&\quad+s^{-1}(sx_1|\cdots|sx_n \otimes 1) - s^{-1}(1 \otimes sx_1|\cdots|sx_n)\\
&\quad+\sum_{j=1}^{n-1}\big[s^{-1}(sx_1|\cdots|sx_j \otimes 1) \cdot (1 \otimes sx_{j+1}|\cdots|sx_n)\\
&\qquad-(-1)^{\theta_j(\theta_n-\theta j)} s^{-1}(1 \otimes sx_{j+1}|\cdots|sx_n) \otimes sx_1|\cdots|sx_j\big]
\end{aligned}$$

with θ_j as above.

Observe that according to this definition, if $c = sx_1|\cdots|sx_n$, then $D(1\otimes c)$ is of the following form.

$$D(1\otimes c) = 1\otimes d_{\mathcal{B}}c + \sum_{i<j} s^{-1}\big(\lambda_{ij,k}(c)\otimes \lambda^k_{ij}(c)\big)\otimes sx_i|\cdots|sx_j$$

where we are applying the Einstein summation convention (cf., section 1.1).

(3) Let $a = sx_1|\cdots|sx_m, b = sy_1|\cdots|sy_n \in \mathcal{B}C^*K$. Then we define

$$\begin{aligned}
&(1\otimes a)(1\otimes b)\\
=&1\otimes a\star b\\
&+\sum\Big[\pm s^{-1}\big(\Theta^\sharp(\lambda_{ij,k}(a)\otimes\lambda_{pq,r}(b))\otimes\lambda^k_{ij}(a)\star\lambda^r_{pq}(b)\big)\\
&\qquad\pm s^{-1}\big(\lambda_{ij,k}(a)\star\lambda_{pq,r}(b)\otimes\Theta^\sharp(\lambda^k_{ij}(a)\otimes\lambda^r_{pq}(b))\big)\\
&\qquad\pm s^{-1}\big(\Theta^\sharp(\lambda_{ij,k}(a)_s\otimes\lambda_{pq,r}(b)_t)\otimes\lambda^k_{ij}(a)_u\star\lambda^r_{pq}(b)_v\big)\cdot\\
&\qquad\qquad\cdot s^{-1}\big(\lambda_{ij,k}(a)^s\star\lambda_{pq,r}(b)^t\otimes\Theta^\sharp(\lambda^k_{ij}(a)^u\otimes\lambda^r_{pq}(b)^v)\big)\Big]\\
&\qquad\qquad\qquad\otimes(sx_i|\cdots|sx_j)\star(sy_p|\cdots|sy_q)
\end{aligned}$$

where we have suppressed the relatively obvious, though horrible to specify, signs given by the Koszul convention (cf., section 1.1). Here $\Theta^\sharp$ is the coderivation homotopy for the commutativity of the multiplication in $\mathcal{B}C^*K$, dual of the derivation homotopy Θ for the cocommutativity of ψ_K, and the extra lower and upper indices in the last summand indicate factors of the coproduct evaluated on the terms in question, e.g., $\Delta(\lambda_{ij,k}(a)) = \lambda_{ij,k}(a)_s\otimes\lambda_{ij,k}(a)^s$.

Observe that, in particular,

$$(1\otimes a)(1\otimes b)\in T^{\leq 2}s^{-1}\mathcal{B}C^*K\otimes\mathcal{B}C^*K.$$

Dupont and Hess showed in [DH4] that the conditions above specify an associative cochain algebra, when the product $\psi^\sharp_K$ is strictly commutative, so that condition (3) reduces to $(1\otimes a)\cdot(1\otimes b) = 1\otimes(a\star b)$. The general, homotopy-commutative case was established by Blanc in his thesis [Bl].

Extend $\varphi:\Omega(\mathcal{B}C^*K\otimes\mathcal{B}C^*K)\longrightarrow\Omega\mathcal{B}C^*K$ to

$$\overline{\varphi}:\Omega(\mathcal{B}C^*K\otimes\mathcal{B}C^*K)\odot\mathcal{B}C^*K\longrightarrow\Omega\mathcal{B}C^*K$$

by $\overline{\varphi}(1\otimes w) = 0$ for all $w\in(\mathcal{B}C^*K)^+$. It is easy to see that $\overline{\varphi}$ is a differential map and that, as explained in [DH2], a straightforward spectral sequence argument shows as well that $\overline{\varphi}$ is a quasi-isomorphism. If $\psi_K^\sharp$ is strictly commutative, then $\overline{\varphi}$ is a strict algebra map, as proved in [DH4]. More generally, Blanc showed in [Bl] that $\overline{\varphi}$ is a quasi-algebra map when $\psi_K^\sharp$ is commutative up to a derivation homotopy. We have thus established the following result.

Lemma 2.2.3. *There is a commuting diagram*

$$\begin{array}{ccc}
\Omega(\mathcal{B}C^*K \otimes \mathcal{B}C^*K) & \xrightarrow{\quad\varphi\quad} & \Omega\mathcal{B}C^*K \\
& {\scriptstyle\iota}\searrow \qquad \nearrow{\scriptstyle\overline{\varphi}}\,{\scriptstyle\simeq} & \\
& \Omega(\mathcal{B}C^*K \otimes \mathcal{B}C^*K) \odot \mathcal{B}C^*K &
\end{array}$$

where ι is the inclusion, $\overline{\varphi}(1 \odot w) = 0$ for all $w \in (\mathcal{B}C^*K)^+$, and $\overline{\varphi}$ is a quasi-algebra quasi-isomorphism.

Now consider the following commutative diagram

$$\begin{array}{ccccc}
\Omega(\mathcal{B}C^*K \otimes \mathcal{B}C^*K) & \xrightarrow[\simeq]{\alpha} & C^*(K \times K) & \xrightarrow{C^*p} & C^*PK \\
\Big\downarrow{\scriptstyle\iota} & & & & \Big\downarrow{\scriptstyle C^*i} \\
\Omega(\mathcal{B}C^*K \otimes \mathcal{B}C^*K) \odot \mathcal{B}C^*K & \xrightarrow[\simeq]{\overline{\varphi}} & \Omega\mathcal{B}C^*K & \xrightarrow[\simeq]{\gamma} & C^*K
\end{array}$$

which satisfies the conditions of Proposition 1.3.7. Hence, we can lift $\overline{\varphi}\gamma$ through C^*i, obtaining a quasi-algebra map

$$\beta : \Omega(\mathcal{B}C^*K \otimes C^*K) \odot \mathcal{B}C^*K \longrightarrow C^*PK$$

such that $\beta\iota = C^*p \circ \alpha$ and $C^*i\beta = \gamma \circ \overline{\varphi}$. We can therefore take

$$\begin{array}{ccc}
\Omega(\mathcal{B}C^*K \otimes \mathcal{B}C^*K) & \xrightarrow{\iota} & \Omega(\mathcal{B}C^*K \otimes \mathcal{B}C^*K) \odot \mathcal{B}C^*K \\
{\scriptstyle\simeq}\Big\downarrow{\scriptstyle\alpha} & & {\scriptstyle\simeq}\Big\downarrow{\scriptstyle\beta} \\
C^*(K \times K) & \xrightarrow{\quad C^*p \quad} & C^*PK
\end{array} \qquad (2.2)$$

as an algebraic model of the path fibration.

2.2.3 The free loop space model

We now twist together the models 2.1 and 2.2, in the spirit of Theorem 1.3.8.

Theorem 2.2.4. [DH4],[Bl] *There is a twisted algebra extension*

$$\bar{\iota} : \Omega\mathcal{B}C^*K \longrightarrow \Omega\mathcal{B}C^*K \odot \mathcal{B}C^*K$$

and a quasi-algebra quasi-isomorphism

$$\delta : \Omega\mathcal{B}C^*K \odot \mathcal{B}C^*K \xrightarrow{\simeq} C^*(\mathcal{L}K)$$

defined as follows.

(1) *The left action of $\Omega\mathcal{B}C^*K$ is specified recursively by*

$$\begin{aligned}(s^{-1}a \otimes 1) \cdot (1 \otimes sx_1|\cdots|sx_n)\\ =& s^{-1}a \otimes sx_1|\cdots|sx_n\\ &- \sum_{j=1}^{n-1}\Big[s^{-1}(a \star (sx_1|\cdots|sx_j)) \cdot (1 \otimes sx_{j+1}|\cdots|sx_n)\\ &\qquad + (-1)^{(\theta_j)(\theta_n-\theta_j)}s^{-1}(a \star (sx_{j+1}|\cdots|sx_n)) \otimes sx_1|\cdots|sx_j\Big]\end{aligned}$$

where $\theta_j = j + \sum_{i\leq j}|x_i|$.

(2) *The restriction of the differential $\overline{D}$ of the twisted extension to $1 \otimes \mathcal{B}C^*K$ is specified recursively by*

$$\begin{aligned}\overline{D}(1 \otimes sx_1|\cdots|sx_n) &= 1 \otimes d_{\mathcal{B}}(sx_1|\cdots|sx_n)\\ &+ \sum_{j=1}^{n-1}[(s^{-1}(sx_1|\cdots|sx_j) \otimes 1) \cdot (1 \otimes sx_{j+1}|\cdots|sx_n)\\ &\qquad - (-1)^{\theta_j(\theta_n-\theta j)}s^{-1}(sx_{j+1}|\cdots|sx_n) \otimes sx_1|\cdots|sx_j]\end{aligned}$$

with θ_j as above.

(3) *Let $a = sx_1|\cdots|sx_m, b = sy_1|\cdots|sy_n \in \mathcal{B}C^*K$. Then we define*

$$\begin{aligned}(1 \otimes a)(1 \otimes b)\\ =& 1 \otimes a \star b\\ &+ \sum(-1)^{\zeta_{ij,pq}}s^{-1}\Big(\Theta^\sharp(\lambda_{ij,k}(a) \star \lambda^k_{ij}(a) \otimes \lambda_{pq,r}(b)) \star \lambda^r_{pq}(b)\Big)\\ &\qquad\qquad \otimes (sx_i|\cdots|sx_j) \star (sy_p|\cdots|sy_q).\end{aligned}$$

Here we are using the same notation as in the definition of the path space model, and

$$\zeta_{ij,pq} = (|\lambda_{pq,r}(b)| + |\lambda^r_{pq}(b)|) \cdot (\theta_j - \theta_{i-1}).$$

(4) *For all* $a, b \in \mathcal{B}C^*K$,

$$\begin{aligned}\delta(s^{-1}a \otimes b) :&= C^*\overline{\Delta} \circ \beta(s^{-1}(a \otimes 1) \otimes b)\\ &= C^*\overline{\Delta} \circ \beta(1 \otimes b) \cdot C^*\overline{p} \circ \gamma(s^{-1}a).\end{aligned}$$

This theorem was proved in the strictly commutative case in [DH4] and in the homotopy-commutative case in [Bl].

In the article [BH] based on part of Blanc's thesis, we use this model to compute the free loop space cohomology algebra of a space X not in the "Anick" range, i.e., where the space does not have a strictly commutative model. This is the first time an explicit calculation of this type has been carried out.

Related work 2.2.5. Our free loop space model construction is patterned on the construction for rational spaces, due to Sullivan and Vigué [SV]. They used their construction to prove the rational *Closed Geodesic Conjecture.*

Kuribayashi, alone [K] and with Yamaguchi [KY], has applied the Eilenberg-Moore spectral sequence to calculating the cohomology algebra of certain free loop spaces.

Let $\mathfrak{C}_*(A)$ denote the Hochschild complex of a (co)chain algebra A. Let N^*X denote the normalized singular cochains on a topological space X with coefficients in a field $\Bbbk$. Ndombol and Thomas have applied SHC-algebra methods to showing that there is a natural map of cochain complexes $\mathfrak{C}_*(N^*X) \longrightarrow C^*(\mathcal{L}X)$ inducing an isomorphism of graded algebras in cohomology. In a similar vein, Menichi proved in [Me2] that $\mathrm{H}^*(\mathcal{L}X)$ is isomorphic to the Hochschild cohomology of the singular chains on ΩX.

Very recently Bökstedt and Ottosen have developed yet another promising method for calculating free loop space cohomology, via a Bousfield spectral sequence [BO1]. Their method, which they have applied explicitly to spaces X with $\mathrm{H}^*(X; \mathbb{F}_2)$ a truncated polynomial algebra on one generator, allows them to obtain the module structure of $\mathrm{H}^*(X; \mathbb{F}_2)$ over the Steenrod algebra.

2.3 The free loop model for topological spaces

In this section we adapt the free loop model above to the case of topological spaces, in order to facilitate construction of a model of the S^1-homotopy orbits on the free loop space. We sacrifice perhaps a bit of the multiplicative structure of the model in section 2.2, but since we are interested in this chapter only in the linear structure of the homotopy orbit cohomology, this is not a great loss.

As usual let X be a 1-connected space with the homotopy type of a finite-type CW-complex, and let K be a finite-type, 1-reduced simplicial set such that $|K| \simeq X$. We again adopt the notation φ for $\Omega\psi_K^\sharp$, where $(\Omega C_*K, \psi_K)$ is the canonical Adams-Hilton model of section 1.2. Recall furthermore that ε_K denotes the unit of the cobar-bar adjunction.

The unit η of the $(|\cdot|, \mathcal{S}_\bullet)$-adjunction induces an injective quasi-isomorphism of chain coalgebras

$$C_*\eta_K : C_*K \xrightarrow{\simeq} S_*|K|$$

that dualizes to a surjective quasi-isomorphism of cochain algebras

$$C^*\eta_K : S^*|K| \xrightarrow{\simeq} C^*K .$$

Furthermore, for all spaces Y, subdivision of cubes defines a quasi-isomorphism of chain coalgebras

$$CU_*Y \xrightarrow{\simeq} S_*Y,$$

which, upon dualization, gives rise to a quasi-isomorphism of cochain algebras

$$S^*Y \xrightarrow{\simeq} CU^*Y.$$

We begin the topological adaptation of the model for the simplicial free loop space by carefully specifying a model of the diagonal map, directly in the cubical cochains, which is where we know how to work explicitly with circle actions, as we show in chapter 3.

Theorem 2.3.1. *There is a noncommutative model of the diagonal map*

$$\begin{array}{ccc}
\Omega(\mathcal{B}C^*K \otimes \mathcal{B}C^*K) & \xrightarrow{\varphi} & \Omega\mathcal{B}C^*K \\
\simeq \big\downarrow \tilde{\alpha} & & \simeq \big\downarrow \tilde{\gamma} \\
CU^*(X \times X) & \xrightarrow{CU^*\Delta} & CU^*X
\end{array}$$

such that

(1) $\tilde{\alpha} \circ \iota_k = C^*pr_k \circ \tilde{\gamma}$ *for* $k = 1, 2$, *where*

$$\iota_k : \Omega\mathcal{B}C^*K \longrightarrow \Omega(\mathcal{B}C^*K \otimes \mathcal{B}C^*K)$$

is the inclusion into the kth tensor factor and pr_k *is the projection onto the kth factor, and*

(2) $\ker \varepsilon_K \subseteq \ker \tilde{\gamma}$.

We do not claim here that $\tilde{\gamma}$ and $\tilde{\alpha}$ are strict algebra maps. In fact, it is probably impossible in general for $\tilde{\gamma}$ simultaneously to be an algebra map and to satisfy $\ker \varepsilon_K \subseteq \ker \tilde{\gamma}$. We show in the proof, however, that $\tilde{\gamma}$ and $\tilde{\alpha}$ are at least quasi-algebra maps.

Proof. Consider the following commutative diagram, in which the solid arrows are given and are all cochain algebra maps. We will explain step-by-step the construction and properties of the dashed arrows.

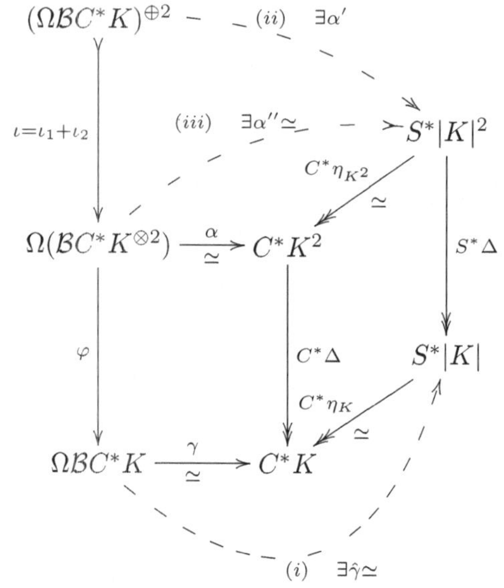

Step (i): Consider the following commutative diagram of cochain complexes.

$$\begin{array}{ccc} \ker\varepsilon_K & \xrightarrow{\quad 0 \quad} & S^*|K| \\ \Big\downarrow{\scriptstyle \text{incl.}} & & \simeq\Big\downarrow{\scriptstyle C^*\eta_K} \\ \Omega\mathcal{B}C^*K & \xrightarrow[\simeq]{\gamma=\varepsilon_K} & C^*K \end{array}$$

Since $C^*\eta_K$ is a surjective quasi-isomorphism and the inclusion map on the left is a free extension of cochain complexes, there exists $\hat{\gamma} : \Omega\mathcal{B}C^*K \longrightarrow S^*|K|$ such that $\ker\varepsilon_K \subset \ker\hat{\gamma}$ and $S^*|K| \circ \hat{\gamma} = \varepsilon_K$. In particular $\hat{\gamma}$ is a quasi-isomorphism by "2-out-of-3" and $\mathrm{H}^*\,\hat{\gamma} = (\mathrm{H}^*\,\eta_K)^{-1} \circ \mathrm{H}^*\,\gamma$ is a map of algebras, i.e., $\hat{\gamma}$ is a quasi-algebra map.

Step (ii): Let $\alpha' = S^*pr_1 \circ \hat{\gamma} + S^*pr_2 \circ \hat{\gamma}$. Since $\varphi\iota_k = Id$ for $k = 1, 2$ and $S^*\Delta \circ S^*pr_k = Id$, the diagram

$$\begin{array}{ccc} (\Omega\mathcal{B}C^*K)^{\oplus 2} & \xrightarrow{\alpha'} & S^*|K|^2 \\ \Big\downarrow{\scriptstyle \varphi\iota} & & \Big\downarrow{\scriptstyle S^*\Delta} \\ \Omega\mathcal{B}C^*K & \xrightarrow{\hat{\gamma}} & S^*|K| \end{array}$$

commutes, by the universal property of the direct sum.

Step (iii): Consider finally the commutative diagram below of cochain complexes and maps.

$$\begin{array}{ccc} (\Omega\mathcal{B}C^*K)^{\oplus 2} & \xrightarrow{\alpha'} & S^*|K|^2 \\ \Big\downarrow{\scriptstyle \iota} & & \simeq \Big\downarrow{\scriptstyle C^*\eta_{K^2}} \\ \Omega(\mathcal{B}C^*K^{\otimes 2}) & \xrightarrow{\alpha} & C^*K^2 \end{array}$$

As usual, since $C^*\eta_{K^2}$ is a surjective quasi-isomorphism and ι is a free extension of cochain complexes, α lifts to $\alpha'' : \Omega(\mathcal{B}C^*K^{\otimes 2}) \longrightarrow S^*|K|^2$, which is a quasi-algebra quasi-isomorphism such that $\alpha''\iota = \alpha'$ and $C^*\eta_{K^2} \circ \alpha'' = \alpha$.

Notice that

$$\begin{aligned} C^*\eta_K \circ S^*\Delta \circ \alpha'' &= C^*\Delta \circ C^*\eta_{K^2} \circ \alpha'' \\ &= C^*\Delta \circ \alpha \\ &= \gamma \circ \varphi \\ &= C^*\eta_K \circ \hat{\gamma} \circ \varphi. \end{aligned}$$

We would like to have $S^*\Delta \circ \alpha'' = \hat{\gamma} \circ \varphi$, but, as we show next, all we can be sure of is that $S^*\Delta \circ \alpha'' \simeq \hat{\gamma} \circ \varphi$, as cochain maps. We then apply the homotopy extension property of free extensions to "fix" α'' and thus conclude the proof.

For any cochain complex, (A, d), with free underlying graded abelian group, let $I(A, d) = (A \oplus A \oplus sA, D)$ denote its canonical cylinder. Let $\mathrm{j}_k : A \longrightarrow I(A, d)$ denote the inclusion onto the kth summand ($k = 1, 2$). Note that j_k is always a quasi-isomorphism.

Let (P, d) be the cochain complex such that the following diagram is a push-out in the category of cochain complexes.

$$\begin{array}{ccc} \Omega(\mathcal{B}C^*K)^{\oplus 2} \oplus \Omega(\mathcal{B}C^*K)^{\oplus 2} & \xrightarrow{\mathrm{j}_1 + \mathrm{j}_2} & I(\Omega(\mathcal{B}C^*K)^{\oplus 2}) \\ \Big\downarrow{\scriptstyle \iota \oplus \iota} & & \Big\downarrow \\ \Omega(\mathcal{B}C^*K^{\otimes 2}) \oplus \Omega(\mathcal{B}C^*K^{\otimes 2}) & \longrightarrow & (P, d) \end{array}$$

We then have a commuting diagram

$$\begin{array}{ccc} (P, d) & \xrightarrow{(S^*\Delta \circ \alpha'' + \hat{\gamma}\varphi) + G} & S^*|K| \\ \Big\downarrow{\scriptstyle \text{incl.}} & & \simeq \Big\downarrow{\scriptstyle C^*\eta_K} \\ I(\Omega(\mathcal{B}C^*K^{\otimes 2})) & \xrightarrow{H} & C^*K \end{array}$$

where G is the constant homotopy (i.e., $G(s\Omega(\mathcal{B}C^*K)^{\oplus 2}) = \{0\}$) from $S^*\Delta \circ \alpha' = \hat{\gamma}\varphi\iota$ to itself and H is the constant homotopy from $C^*\Delta \circ \alpha = \gamma\varphi$ to itself. Again, since $C^*\eta_K$ is a surjective quasi-isomorphism and the inclusion on the left is a free extension, we can lift the homotopy H to

$$H' : I\big(\Omega(\mathcal{B}C^*K^{\otimes 2})\big) \longrightarrow S^*|K|$$

such that

$$H'|_{(P,d)} = (S^*\Delta \circ \alpha'' + \hat{\gamma}\varphi) + G,$$

i.e., H' is a homotopy from $S^*\Delta \circ \alpha''$ to $\hat{\gamma}\varphi$ that is constant on the subcomplex $\Omega(\mathcal{B}C^*K)^{\oplus 2}$.

Let $G' : I(\Omega(\mathcal{B}C^*K)^{\oplus 2}) \longrightarrow S^*|K|^2$ be the constant homotopy from α' to itself. In particular, $S^*\Delta \circ G' = G$. Consider the push-out diagram of cochain complexes

$$\begin{array}{ccc} \Omega(\mathcal{B}C^*K)^{\oplus 2} & \overset{\mathrm{j}_1}{\underset{\simeq}{\rightarrowtail}} & I(\Omega(\mathcal{B}C^*K)^{\oplus 2}) \\ \downarrow \iota & & \downarrow \\ \Omega(\mathcal{B}C^*K^{\otimes 2}) & \overset{\hat{\mathrm{j}}_1}{\underset{\simeq}{\rightarrowtail}} & (Q,d) \end{array}$$

$$G' : I(\Omega(\mathcal{B}C^*K)^{\oplus 2}) \to S^*|K|^2, \quad \alpha''+G' : (Q,d) \overset{\simeq}{\dashrightarrow} S^*|K|^2, \quad \alpha'' : \Omega(\mathcal{B}C^*K^{\otimes 2}) \overset{\simeq}{\longrightarrow} S^*|K|^2$$

in which all arrows are free extensions, since the push-out of a cofibration is a cofibration. Furthermore, $\hat{\mathrm{j}}_1$ is a quasi-isomorphism because the push-out of an acyclic cofibration is an acyclic cofibration. By "2-out-of-3", $\alpha'' + G'$ is a quasi-isomorphism.

Let $\mathfrak{k} : I(\Omega(\mathcal{B}C^*K)^{\oplus 2}) \rightarrowtail I\big(\Omega(\mathcal{B}C^*K^{\otimes 2})\big)$ be the natural inclusion. The triangle

$$\Omega(\mathcal{B}C^*K^{\otimes 2}) \overset{\hat{\mathrm{j}}_1}{\underset{\simeq}{\rightarrowtail}} (Q,d) \overset{\mathrm{j}_1+\mathfrak{k}}{\rightarrowtail} I\big(\Omega(\mathcal{B}C^*K^{\otimes 2})\big)$$

$$\mathrm{j}_1 : \Omega(\mathcal{B}C^*K^{\otimes 2}) \overset{\simeq}{\longrightarrow} I\big(\Omega(\mathcal{B}C^*K^{\otimes 2})\big)$$

then commutes, proving that $\mathrm{j}_1 + \mathfrak{k}$ is also a quasi-isomorphism.

We now have a commuting diagram

$$\begin{array}{ccc} (Q,d) & \overset{\alpha''+G'}{\underset{\simeq}{\longrightarrow}} & S^*|K|^2 \\ \simeq \big\downarrow \mathrm{j}_1+\mathfrak{k} & & \big\downarrow S^*\Delta \\ I\big(\Omega(\mathcal{B}C^*K^{\otimes 2})\big) & \overset{H'}{\longrightarrow} & S^*|K| \end{array}$$

and therefore, since $\mathrm{j}_1 + \mathfrak{k}$ is a free extension and a quasi-isomorphism and $S^*\Delta$ is surjective, there is a homotopy $\hat{H} : I\big(\Omega(\mathcal{B}C^*K^{\otimes 2})\big) \longrightarrow S^*|K|^2$ such that

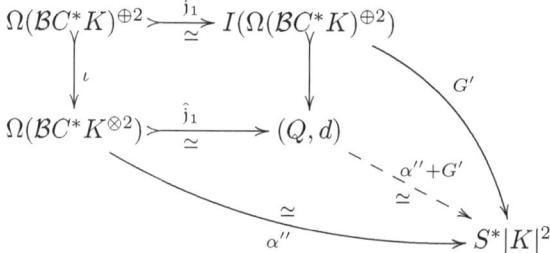

$\widehat{H}\circ(\mathfrak{j}_1+\mathfrak{k})=\alpha''+G'$ and $S^*\Delta\circ\widehat{H}=H'$. Furthermore, $\widehat{H}$ is a quasi-isomorphism by "2-out-of-3".

Let $\hat{\alpha}=\widehat{H}\circ\mathfrak{j}_2$. Then $\hat{\alpha}$ is a quasi-algebra quasi-isomorphism, as it is homotopic to a quasi-algebra quasi-isomorphism. Moreover,

$$\hat{\alpha}\iota=\widehat{H}\mathfrak{k}\mathfrak{j}_2=G'\mathfrak{j}_2=\alpha'=S^*pr_1\circ\hat{\gamma}+S^*pr_2\circ\hat{\gamma},$$

while

$$S^*\Delta\circ\hat{\alpha}=H'\mathfrak{j}_2=\hat{\gamma}\varphi,$$

i.e, the following diagram commutes exactly and is therefore a noncommutative model.

$$\begin{array}{ccc}\Omega(\mathcal{B}C^*K\otimes\mathcal{B}C^*K) & \xrightarrow{\varphi} & \Omega\mathcal{B}C^*K\\ \simeq\big\downarrow\hat{\alpha} & & \simeq\big\downarrow\hat{\gamma}\\ S^*|K|^2 & \xrightarrow{S^*\Delta} & S^*|K|\end{array}$$

To complete the proof, compose $\hat{\alpha}$ and $\hat{\gamma}$ with the natural cochain algebra quasi-isomorphisms

$$S^*|K|^2\xrightarrow{\simeq}CU^*X^2\quad\text{and}\quad S^*|K|\xrightarrow{\simeq}CU*X$$

to obtain $\tilde{\alpha}$ and $\tilde{\gamma}$. □

The next step in our simplification of the free loop model is to find an appropriate model for the topological path fibration

$$p:X^I\twoheadrightarrow X\times X:\ell\longmapsto(\ell(0),\ell(1))\,.$$

The advantage to working directly with cubical cochains is that we are able to define an explicit model morphism, extending $\tilde{\alpha}$, from $\Omega(\mathcal{B}C^*K\otimes\mathcal{B}C^*K)\odot\mathcal{B}C^*K$ (defined exactly as in section 2.2) to CU^*X^I. This may be possible simplicially as well, but the formulas are most probably not nearly as simple.

Given $x\in X$, let ℓ_x denote the constant path at x. Let $i:X\longrightarrow X^I$ be defined by $i(x)=\ell_x$. Let $H:CU_*X^I\longrightarrow CU_{*+1}X^I$ be the chain homotopy such that for all $T\in CU_nX^I$,

$$H(T)(t_0,...,t_n):=T(t_1,...,t_n)(t_0\cdot -):I\longrightarrow X.$$

It is easy to see that $[d,H](T)=\ell_{T(-,...,-)(0)}-T$, i.e., $[d,H]=\iota\pi-Id$, where

$$\iota:\operatorname{Im}CU_*i\longrightarrow CU_*X^I$$

is the inclusion and

$$\pi:CU_*X^I\longrightarrow\operatorname{Im}CU_*i$$

is defined by $\pi(T) = \ell_{T(-,...,-)(0)}$. Observe that $H\iota = 0$, as $H(T)$ is degenerate if the path $T(t_1, ..., t_n)$ is constant for all $(t_1, ..., t_n)$. The homotopy H therefore induces a homotopy

$$H' : \operatorname{coker} CU_* i \longrightarrow \operatorname{coker} CU_{*+1} i : [T] \longmapsto [H(T)]$$

satisfying $[d, H'] = -Id$, i.e., H' is a contracting homotopy.

Upon dualizing, we obtain a cochain homotopy

$$J = -(H')^\sharp : \ker CU^* i \longrightarrow \ker CU^{*-1} i$$

such that $[d^\sharp, J] = Id$. Note that $[d^\sharp, J] = Id$ implies that $d^\sharp J^2 = J^2 d^\sharp$.

It is important for the constructions in chapter 3 to observe as well that J is a $(0, Id)$-derivation, i.e.,

$$J(fg) = J(f) \cdot g$$

for all $f, g \in \ker CU^* i$, since $\overline{\Delta} H = (\iota\pi \otimes H - H \otimes Id)\overline{\Delta}$, which implies that $\overline{\Delta} H' = -(H' \otimes Id)\overline{\Delta}$, where $\overline{\Delta}$ denotes the reduced coproduct on $CU_* X^I$ or $\operatorname{coker} CU_* i$.

We resume the construction above in the following proposition.

Proposition 2.3.2. *Let* $i : X \longrightarrow X^I$ *be defined by* $i(x) = \ell_x$, *the constant path at* x. *There is a natural cochain homotopy*

$$J : \ker CU^* i \longrightarrow \ker CU^{*-1} i$$

such that

(1) $[d^\sharp, J] = Id$,

(2) $d^\sharp J^2 = J^2 d^\sharp$, *and*

(3) $J(fg) = J(f) \cdot g$.

We now apply J to giving an explicit definition of a model morphism $\tilde{\beta}$ from $\Omega(\mathcal{B}C^* K \otimes \mathcal{B}C^* K) \odot \mathcal{B}C^* K$ to $CU^* X^I$.

Proposition 2.3.3. *There is a noncommutative model*

$$\begin{array}{ccc} \Omega(\mathcal{B}C^*K \otimes \mathcal{B}C^*K) & \xrightarrow{\ \iota\ } & \Omega(\mathcal{B}C^*K \otimes \mathcal{B}C^*K) \odot \mathcal{B}C^*K \\ \simeq \downarrow \tilde{\alpha} & & \simeq \downarrow \tilde{\beta} \\ CU^*(X \times X) & \xrightarrow{\ CU^*p\ } & CU^* X^I \end{array}$$

where $\tilde{\beta}$ *is defined recursively by*

$$\tilde{\beta}(w \otimes c) = (-1)^{wc} \tilde{\beta}(1 \otimes c) \cdot CU^* p \circ \tilde{\alpha}(w)$$

and

$$\tilde{\beta}(1 \otimes c) = J\tilde{\beta}D(1 \otimes c).$$

Proof. Suppose that if $\tilde{\beta}$ is defined in accordance with the formulas in the statement above on $\Omega(\mathcal{B}C^*K \otimes \mathcal{B}C^*K) \odot (\mathcal{B}C^*K)_{<n}$, then $d^\sharp\tilde{\beta} = \tilde{\beta}D$ and $CU^*i \circ \tilde{\beta} = \tilde{\gamma} \circ \bar{\varphi}$ when restricted to this same complex.

Let $c \in (\mathcal{B}C^*K)_n$. It is clear from the definition of D that $\tilde{\gamma}\bar{\varphi}D(1 \otimes c) = 0$. Thus, by the induction hypothesis, $\tilde{\beta}D(1 \otimes c) \in \ker CU^*i$, which implies that $J\tilde{\beta}D(1 \otimes c)$ is defined. Moreover,

$$\begin{aligned} d^\sharp J\tilde{\beta}D(1 \otimes c) &= \tilde{\beta}D(1 \otimes c) - Jd^\sharp\tilde{\beta}D(1 \otimes c) \\ &= \tilde{\beta}D(1 \otimes c) - J\tilde{\beta}D^2(1 \otimes c) \\ &= \tilde{\beta}D(1 \otimes c) \end{aligned}$$

and $CU^*iJ\tilde{\beta}D(1 \otimes c) = 0 = \tilde{\gamma}\varphi(1 \otimes c)$, so we can set

$$\tilde{\beta}(1 \otimes c) = J\tilde{\beta}D(1 \otimes c).$$

For the usual reasons, $\tilde{\beta}$ is then a quasi-algebra quasi-isomorphism. □

We can now twist together the models of Theorem 2.3.1 and of Proposition 2.3.3, obtaining a noncommutative model

$$\begin{array}{ccc} \Omega\mathcal{B}C^*K & \longrightarrow & \Omega\mathcal{B}C^*K \odot \mathcal{B}C^*K \\ \simeq\big\downarrow\tilde{\gamma} & & \simeq\big\downarrow\tilde{\delta} \\ CU^*X & \xrightarrow{CU^*e} & CU^*\mathcal{L}X \end{array}$$

where $\Omega\mathcal{B}C^*K \odot \mathcal{B}C^*K$ is defined exactly as in section 2.2 and

$$\tilde{\delta}(w \otimes c) := (-1)^{wc}\tilde{\delta}(1 \otimes c) \cdot CU^*e \circ \tilde{\gamma}(w)$$

and

$$\tilde{\delta}(1 \otimes c) := CU^*jJ\tilde{\beta}D(1 \otimes c).$$

Recall that $j : \mathcal{L}X \longrightarrow X^I$ is the natural inclusion and that $e : \mathcal{L}X \twoheadrightarrow X$ is the basepoint evaluation.

An easy Zeeman's comparison theorem argument shows that $\tilde{\delta}$ is a quasi-isomorphism. Furthermore, applying the Eilenberg-Moore spectral sequence of algebras, one obtains that $\tilde{\delta}$ induces an algebra isomorphism on the E_∞-terms, which is not quite as strong as saying that it is a quasi-algebra map, but is good enough for our purposes in chapter 3. By arguments similar to those in [DH4], we can show that $\tilde{\delta}$ is truly a quasi-algebra map in certain special cases, and it may perhaps be a quasi-algebra map in general.

2.4 Linearization of the free loop model

In this section we simplify even further the free loop model, making it as small as possible, to facilitate homotopy orbit space computations in chapter 3.

Consider the surjection

$$\varepsilon_K \otimes Id : \Omega\mathcal{B}C^*K \otimes \mathcal{B}C^*K \longrightarrow C^*K \otimes \mathcal{B}C^*K.$$

Extend the differential on C^*K to a differential $\check{D}$ on $C^*K \otimes \mathcal{B}C^*K$, which is a free right C^*K-module, by

$$\check{D}(1 \otimes c) := (\varepsilon_K \otimes Id)\overline{D}(1 \otimes c),$$

extended as a right module derivation. Consequently, $\check{D}(\varepsilon_K \otimes Id) = (\varepsilon_K \otimes Id)\overline{D}$, which implies in turn that $\check{D}^2 = 0$.

Let $\pi : \mathcal{B}C^*K \longrightarrow C^*K$ be the projection onto (desuspended) linear terms, i.e.,

$$\pi(sx_1|\cdots|sx_n) := \begin{cases} x_1 : n = 1 \\ 0 : n > 1. \end{cases}$$

Using the notation introduced in the definition of $\Omega(\mathcal{B}C^*K \otimes \mathcal{B}C^*K) \odot \mathcal{B}C^*K$ in section 2.2, we obtain the following explicit formula for $\check{D}$, when $c = sx_1|\cdots|sx_n$.

$$\check{D}(1 \otimes c) = 1 \otimes d_{\mathcal{B}}c + \sum_{i<j} \pi\big(\lambda_{ij,k}(c) \star \lambda^k_{ij}(c)\big) \otimes sx_i|\cdots|sx_j.$$

Notice that it is entirely possible that $\check{D}(1 \otimes c) \neq 1 \otimes d_{\mathcal{B}}c$, since the formula for ψ_K implies that if $\deg x \gg 0$, then $\psi_K(s^{-1}x)$ has a nonzero summand in $T^{>1}s^{-1}C_+K \otimes s^{-1}C_+K$.

Define now a left C^*K-action on $C^*K \otimes \mathcal{B}C^*K$ by

$$(x \otimes 1) \cdot (1 \otimes c) := (\varepsilon_K \otimes Id)\big((s^{-1}(sx) \otimes 1) \cdot (1 \otimes c)\big)$$

for all $x \in C^*K$ and $c \in \mathcal{B}C^*K$. If $\overline{D}(1 \otimes c) - 1 \otimes d_{\mathcal{B}}c = \sum_i s^{-1}(a_i) \otimes b^i$, then

$$\begin{aligned}
&\check{D}\big((x \otimes 1) \cdot (1 \otimes c)\big) \\
=&(\varepsilon_K \otimes Id)\big((s^{-1}(sdx) \otimes 1) \cdot (1 \otimes c) + (-1)^x (s^{-1}(sx) \otimes 1) \cdot \overline{D}(1 \otimes c)\big) \\
=&(dx \otimes 1) \cdot (1 \otimes c) \\
&+ (-1)^x (\varepsilon_K \otimes Id)\Bigg((s^{-1}(sx) \otimes 1) \cdot \big(1 \otimes d_{\mathcal{B}}c \\
&\qquad\qquad + \sum_i (-1)^{(a_i+1)b^i} (1 \otimes b^i)(s^{-1}(sa_i) \otimes 1)\big)\Bigg) \\
=&(dx \otimes 1) \cdot (1 \otimes c) \\
&+ (-1)^x \Bigg((x \otimes 1) \cdot \big(1 \otimes d_{\mathcal{B}}c + \sum_i (-1)^{(a_i+1)b^i} (1 \otimes b^i)(a_i \otimes 1)\big)\Bigg) \\
=&(dx \otimes 1) \cdot (1 \otimes c) + (-1)^x (x \otimes 1) \cdot \check{D}(1 \otimes c),
\end{aligned}$$

i.e., the left C^*K action commutes with the differential.

Again using the notation of section 2.2, we can write

$$(x \otimes 1) \cdot (1 \otimes c) = x \otimes c - \sum_{i<j} \pi\big(x \star \lambda_{ij,k}(c) \star \lambda^k_{ij}(c)\big) \otimes sx_i|\cdots|sx_j.$$

The following proposition summarizes the observations above. Let $C^*K\widetilde{\otimes}\mathcal{B}C^*K$ denote $C^*K \otimes \mathcal{B}C^*K$ endowed with the differential $\check{D}$ and the C^*K-bimodule structure defined above.

Proposition 2.4.1. *There is a twisted bimodule extension*

$$C^*K \longrightarrow C^*K\widetilde{\otimes}\mathcal{B}C^*K \longrightarrow \mathcal{B}C^*K$$

such that $\varepsilon_K \otimes Id : \Omega\mathcal{B}C^*K\widetilde{\otimes}\mathcal{B}C^*K \longrightarrow C^*K\widetilde{\otimes}\mathcal{B}C^*K$ *is a map of differential right C^*K-modules. In particular, $\varepsilon_K \otimes Id$ is a quasi-isomorphism.*

Note that $\varepsilon_K \otimes Id$ is not a bimodule map itself, since it is possible that for some $w \in \perp^{>1} sC^+K$ and $c \in \mathcal{B}C^*K$, the product $(s^{-1}(w)\otimes 1)(1\otimes c)$ has a nonzero summand in $s^{-1}(sC^*K)\otimes\mathcal{B}C^*K$, i.e., that $(\varepsilon_K\otimes Id)\big((s^{-1}(w)\otimes 1)(1\otimes c)\big) \neq 0$, even though $s^{-1}(w) \in \ker \varepsilon_K$. On the other hand if we filter both $\Omega\mathcal{B}C^*K \odot \mathcal{B}C^*K$ and $C^*K\widetilde{\otimes}\mathcal{B}C^*K$ by degree in the left tensor factor, then $\varepsilon_K \otimes Id$ induces an isomorphism of bigraded bimodules on the E_∞-terms of the associated spectral sequences, so $\varepsilon_K \otimes Id$ is almost a quasi-bimodule map. It may even be possible to define explicitly a cochain homotopy ensuring that $\varepsilon_K \otimes Id$ truly is a quasi-bimodule map.

For the constructions in chapters 3 and 4, we need a quasi-isomorphism

$$\Upsilon : C^*K\widetilde{\otimes}\mathcal{B}C^*K \xrightarrow{\simeq} CU^*\mathcal{L}X\ ,$$

which we obtain as follows. Recall that ε_K has a differential, though not multiplicative, section

$$\sigma_K : C^*K \longrightarrow \Omega\mathcal{B}C^*K : x \longmapsto s^{-1}(sx).$$

Consider the following commutative diagram of cochain complexes and maps.

$$\begin{array}{ccccc}
C^*K\widetilde{\otimes} \perp^{\leq 2} sC^+K & \xrightarrow{\sigma_K\otimes Id} & \Omega\mathcal{B}C^*K\odot \perp^{\leq 2} sC^+K & \xrightarrow{\text{incl.}} & \Omega\mathcal{B}C^*K \odot \mathcal{B}C^*K \\
\Big\downarrow{\scriptstyle\text{incl.}} & & & & \simeq\Big\downarrow{\scriptstyle\varepsilon_K\otimes Id} \\
C^*K\widetilde{\otimes}\mathcal{B}C^*K & & \xrightarrow{\qquad\qquad Id\qquad\qquad} & & C^*K\widetilde{\otimes}\mathcal{B}C^*K
\end{array}$$

Since the inclusion map on the left is a free extension of cochain complexes and $\varepsilon_K \otimes Id$ is a surjective quasi-isomorphism, we can extend $\sigma_K \otimes Id$ to a cochain map

$$\hat{\sigma} : C^*K\widetilde{\otimes}\mathcal{B}C^*K \longrightarrow \Omega\mathcal{B}C^*K \odot \mathcal{B}C^*K$$

such that $(\varepsilon_K \otimes Id)\hat{\sigma}_K = Id$, i.e., $\hat{\sigma}$ is a section of $\varepsilon_K \otimes Id$. In particular, $\hat{\sigma}$ is a quasi-isomorphism and for all $x \otimes c \in C^*K \otimes \mathcal{B}C^*K$

$$s^{-1}(sx) \otimes c - \hat{\sigma}(x \otimes c) \in \ker \varepsilon_K \otimes \mathcal{B}C^*K. \tag{2.3}$$

Furthermore, like $\varepsilon_K \otimes Id$, $\hat{\sigma}$ induces an isomorphism of bigraded bimodules on the E_∞-terms of the usual spectral sequences and is a quasi-bimodule map if and only if $\varepsilon_K \otimes Id$ is.

We now define Υ to be the composition below.

$$\begin{array}{c} C^*K\widetilde{\otimes}\mathcal{B}C^*K \xrightarrow{\hat{\sigma}} \Omega\mathcal{B}C^*K \odot \mathcal{B}C^*K \xrightarrow{\tilde{\delta}} CU^*\mathcal{L}X \\ \underbrace{\hspace{20em}}_{\Upsilon} \end{array}$$

Observe that 2.3 implies that for all $x \otimes c \in C^*K\widetilde{\otimes}\mathcal{B}C^*K$,

$$\Upsilon(x \otimes c) = \tilde{\delta}(s^{-1}(sx) \otimes c) \tag{2.4}$$

since $\ker \varepsilon_K \subseteq \ker \tilde{\gamma}$.

Definition 2.4.2. Let X be a 1-connected space with the homotopy type of a finite-type CW-complex, and let K be a finite-type, 1-reduced simplicial set such that $|K| \simeq X$. The twisted C^*K-bimodule extension

$$fls^*(X) := C^*K\widetilde{\otimes}\mathcal{B}C^*K$$

together with the quasi-isomorphism

$$\Upsilon : fls^*(X) \xrightarrow{\simeq} CU^*\mathcal{L}X$$

is a *thin free loop model* for X.

We conclude this section with an important observation concerning the relation between Υ and the product $\psi_K^\sharp$.

Proposition 2.4.3. *Suppose that $\psi_K^\sharp$, the multiplication on $\mathcal{B}C^*K$, is commutative. If $x \in C^*K$ is a cycle, then $\Upsilon(1 \otimes c \star sx) = \Upsilon(1 \otimes c) \cdot \Upsilon(1 \otimes sx)$. In particular*

$$\Upsilon(1 \otimes sx_1 \star \cdots \star sx_n) = \Upsilon(1 \otimes sx_1) \cdots \Upsilon(1 \otimes sx_n)$$

if $dx_i = 0$ for some i.

Consequently, if $\psi_K^\sharp$ is commutative, then the commutator $[\Upsilon(1 \otimes sx), \Upsilon(1 \otimes sy)] = 0$, for all cycles $x, y \in C^*K$.

Proof. Recall from section 2.2 that if $\psi_K^\sharp$ is commutative, then we can choose the multiplication in the model $\Omega(\mathcal{B}C^*K \otimes \mathcal{B}C^*K) \odot \mathcal{B}C^*K$ so that $(1 \otimes c) \cdot (1 \otimes c') = 1 \otimes c \star c'$ for all $c, c' \in C^*K$. According to Proposition 2.3.3 we then have that

$$\begin{aligned}
\tilde{\beta}(1 \otimes c \star c') =& J\tilde{\beta}D(1 \otimes c \star c') \\
=& J\tilde{\beta}D\big((1 \otimes c)(1 \otimes c')\big) \\
=& J\tilde{\beta}\big(D(1 \otimes c) \cdot (1 \otimes c') + (-1)^c (1 \otimes c) \cdot D(1 \otimes c')\big) \\
=& J\tilde{\beta}D(1 \otimes c) \cdot \tilde{\beta}(1 \otimes c') + (-1)^c J\tilde{\beta}(1 \otimes c) \cdot \tilde{\beta}D(1 \otimes c') \\
=& \tilde{\beta}(1 \otimes c) \cdot \tilde{\beta}(1 \otimes c') + (-1)^c J^2 \tilde{\beta}D(1 \otimes c) \cdot d^\sharp \tilde{\beta}(1 \otimes c').
\end{aligned}$$

Thus, since $d^\sharp J^2 = J^2 d^\sharp$,

$$\begin{aligned}
\Upsilon(1 \otimes c \star c') =& CU^* j\tilde{\beta}(1 \otimes c \star c') \\
=& \Upsilon(1 \otimes c) \cdot \Upsilon(1 \otimes c') + (-1)^c d^\sharp CU^* jJ^2 \tilde{\beta}(1 \otimes c) \cdot d^\sharp \Upsilon(1 \otimes c) \\
=& \Upsilon(1 \otimes c) \cdot \Upsilon(1 \otimes c') + (-1)^c d^\sharp CU^* jJ^2 \tilde{\beta}(1 \otimes c) \cdot \Upsilon\check{D}(1 \otimes c).
\end{aligned}$$

Thus, if $x \in C^*K$ is a cycle, then $\check{D}(1 \otimes sx) = 0$ and so $\Upsilon(1 \otimes c \star sx) = \Upsilon(1 \otimes c) \cdot \Upsilon(1 \otimes sx)$. The second part of the statement follows by induction. □

Chapter 3

Homotopy orbit spaces

In this section we construct a noncommutative model $hos^*(X)$ for the homotopy orbit space $(\mathcal{L}X)_{hS^1}$ of the natural S^1-action on the free loop space $\mathcal{L}X$. The form of $hos^*(X)$ is, not surprisingly, similar to that of the complex that gives the cyclic cohomology of an algebra. The author is grateful to Nicolas Dupont for the ideas he contributed during our discussions of $(\mathcal{L}X)_{hS^1}$ over the years.

We begin by proving the existence of a very special family of primitive elements in the reduced cubical chains on S^1 and then studying its properties. We then introduce a particularly useful resolution of the cubical chains on ES^1 as a module over the cubical chains on S^1, which we apply to constructing a model of the homotopy orbit space of any S^1-action. In the final part of this section we specialize to the case of $\mathcal{L}X$, obtaining a small, noncommutative model for $(\mathcal{L}X)_{hS^1}$ as an extension of the thin free loop model $fls^*(X)$.

3.1 A special family of primitives

Let $CU_*(X)$ denote the reduced cubical chains on a topological space X. We begin by defining a suspension-type degree $+1$ operation on CU_*S^1.

Definition 3.1.1. If $T : I^n \longrightarrow S^1$ is an n-cube, let $\sigma(T)$ be the $(n+1)$-cube defined by

$$\sigma(T)(t_0, ..., t_n) := \big(T(t_1, ..., t_n)\big)^{t_0},$$

where we are considering S^1 as the unit circle in the complex plane, i.e.,

$$S^1 = \{e^{i\theta} \mid 0 \leq \theta \leq 2\pi\}.$$

Remark 3.1.2. It is clear that $\sigma(T)$ is degenerate if T is degenerate. The operation σ can therefore be extended linearly to all of CU_*S^1.

As the next lemma states, σ is a contracting homotopy in degrees greater than one and is a $(Id, 0)$-coderivation.

Lemma 3.1.3. *Let $T \in CU_*S^1$.*

(1) *If $\deg T \geq 2$, then $d\sigma(T) = T - \sigma(dT)$ where d is the usual differential on CU_*S^1.*

(2) *$\overline{\Delta}(\sigma(T)) = \sigma(T_i) \otimes T^i$, where $\overline{\Delta}$ is the usual reduced coproduct on CU_*S^1 and $\Delta(T) = T_i \otimes T^i$ (using the Einstein summation convention).*

Simple calculations, applying the definitions of the cubical differential and the cubical coproduct, as given for example in [Mas] and [An], suffice to prove this lemma.

We now apply the σ operation to the recursive construction of an important family of elements in CU_*S^1.

Definition 3.1.4. Let $T_0 : I \longrightarrow S^1$ be the 1-cube defined by $T_0(t) = e^{i2\pi t}$. Given $T_k \in CU_{2k+1}S^1$ for all $k < n$, let T_n be the $(2n+1)$-cubical chain defined by

$$T_n := \sigma\Big(\sum_{i=1}^{n} T_{i-1} \cdot T_{n-i}\Big) \in CU_{2n+1}S^1.$$

Let $\mathcal{T} := \{T_n \mid n \geq 0\}$.

Example 3.1.5. It is easy to see that

$$T_1(t_0, t_1, t_2) = e^{i2\pi t_0(t_1+t_2)}$$

and that $T_2 = U + V$ where

$$U(t_0, ..., t_4) = e^{i2\pi t_0(t_1+(t_2+t_3)t_4)} \text{ and } V(t_0, ..., t_4) = e^{i2\pi t_0(t_1(t_2+t_3)+t_4)}.$$

Proposition 3.1.6. *The family $\mathcal{T}$ satisfies the following properties.*

(1) *$dT_0 = 0$ and $0 \neq [T_0]$ in H_1S^1.*

(2) *$dT_n = \sum_{i=1}^{n} T_{i-1}T_{n-i}$ for all $n > 0$.*

(3) *Every T_n is primitive in CU_*S^1.*

Proof. Points (1) and (2) are easy consequences of Lemma 3.1.3. It is well known that T_0 represents the unique nonzero homology generator of H_*S^1.

An easy inductive argument applying Lemma 3.1.3(2) proves point (3), since if T_k is primitive for all $k < n$, then the sum $\sum_{i=1}^{n} T_{i-1} \cdot T_{n-i}$ is also primitive, as it is symmetric and all factors are of odd degree. □

Let $< \mathcal{T} >$ denote the subalgebra of CU_*S^1 generated by the family $\mathcal{T}$. Since all the T_n's are primitive, $< \mathcal{T} >$ is a sub Hopf algebra of CU_*S^1. Proposition 3.1.6(1) and (2) imply that $< \mathcal{T} >$ is closed under the differential, and that the inclusion

$$< \mathcal{T} > \xhookrightarrow{\simeq} CU_*S^1$$

is a quasi-isomorphism.

3.2 A useful resolution of CU_*ES^1

We now put the family $\mathcal{T}$ to work in constructing a simple, neat resolution of CU_*ES^1 as a CU_*S^1-module. To understand why this is important, recall that a special case of Moore's theorem (cf., [Mc], Thm. 7.27) states that for any left S^1-space X and any free CU_*S^1-resolution (Q,d) of CU_*ES^1, there is a diagram of quasi-isomorphisms of chain complexes

$$\begin{array}{ccc} (Q,d)\otimes_{CU_*S^1} CU_*X & \xrightarrow{\simeq} & CU_*(X_{hS^1}) \\ \downarrow{\scriptstyle \pi} & & \downarrow{\scriptstyle \pi} \\ (Q,d)\otimes_{CU_*S^1} \mathbb{Z} & \xrightarrow{\simeq} & CU_*(BS^1) \end{array}$$

where the projection maps π are induced by the map $X \longrightarrow *$. Hence,

$$\operatorname{Tor}_*^{CU_*S^1}(CU_*ES^1, CU_*X) \cong H_*(X_{hS^1}),$$

and so a resolution of CU_*ES^1 provides us with a general tool for computing $H_*(X_{hS^1})$ for an arbitrary S^1-space X.

Let Γ denote the divided powers algebra functor. Recall that if w is in even degree, then

$$\Gamma w = \bigoplus_{k\geq 0} \mathbb{Z}\cdot w(k),$$

where $\deg w(k) = k\cdot |w|$, $w(0) = 1$, $w(1) = w$ and $w(k)w(l) = \binom{k+l}{k} w(k+l)$. Furthermore, Γw is in fact a Hopf algebra, where the coproduct is specified by $\Delta(w) = w\otimes 1 + 1\otimes w$.

Consider $(\Lambda u, 0) = (\mathbb{Z}\cdot u, 0)$, where u is in degree 1, and its acyclic extension $(\Gamma v\otimes \Lambda u, \partial)$, where v is in degree 2 and $\partial v(k) := v(k-1)\otimes u$ for all $k\geq 1$. There is a chain algebra quasi-isomorphism

$$\xi : CU_*S^1 \xrightarrow{\simeq} (\Lambda u, 0)$$

defined by $\xi(T_0) = u$ and $\xi(T) = 0$ for all other cubes T.

Define a semifree extension of right CU_*S^1-modules

$$\iota : CU_*S^1 \longrightarrow (\Gamma v\otimes CU_*S^1, \tilde\partial)$$

by

$$\tilde\partial(v(n)\otimes 1) := \sum_{k=0}^{n-1} (-1)^k v(n-k-1)\otimes T_k.$$

It is an immediate consequence of Proposition 3.1.6(2) that $\tilde\partial^2 = 0$. Furthermore, for all $n\geq 1$,

$$(Id\otimes\xi)\tilde\partial(v(n)\otimes 1) = v(n-1)\otimes u = \partial(Id\otimes\xi)(v(n)\otimes 1),$$

which implies that the CU_*S^1-module map

$$Id \otimes \xi : (\Gamma v \otimes CU_*S^1, \tilde{\partial}) \longrightarrow (\Gamma v \otimes \Lambda u, \partial)$$

is a differential map. A quick Zeeman's Comparison Theorem argument then shows that $Id \otimes \xi$ is a quasi-isomorphism, so that $(\Gamma v \otimes CU_*S^1, \tilde{\partial})$ is acyclic.

We claim that $(\Gamma v \otimes CU_*S^1, \tilde{\partial})$ is a CU_*S^1-resolution of CU_*ES^1. To verify this, observe that there is a commutative diagram of CU_*S^1-modules

$$
\begin{array}{ccc}
CU_*S^1 & \xrightarrow{CU_*j} & CU_*ES^1 \\
\downarrow{\iota} & & \simeq\downarrow \\
(\Gamma v \otimes CU_*S^1, \tilde{\partial}) & \xrightarrow{\simeq} & \mathbb{Z}
\end{array}
$$

where j is the inclusion of S^1 as the base of the construction of ES^1, which is an S^1-equivariant map. Since $(\Gamma v \otimes CU_*S^1, \tilde{\partial})$ is a semifree extension and the map from CU_*ES^1 to $\mathbb{Z}$ is a surjective quasi-isomorphism, we can extend CU_*j to a CU_*S^1-module map

$$\varepsilon : (\Gamma v \otimes CU_*S^1, \tilde{\partial}) \longrightarrow CU_*ES^1$$

which is a quasi-isomorphism by "2-out-of-3".

3.3   Modeling S^1-homotopy orbits

Let X be any (left) S^1-space, where $g : S^1 \times X \longrightarrow X$ is the action map. There is then a natural CU_*S^1 module structure on CU_*X, given by the composition

$$CU_*S^1 \otimes CU_*X \underbrace{\xrightarrow{EZ}_{\simeq} CU_*(S^1 \times X) \xrightarrow{CU_*g} CU_*X}_{\kappa}$$

where EZ denotes the Eilenberg-Zilber equivalence. Observe that κ is a coalgebra map, as it is the composition of two coalgebra maps. Since Moore's Theorem implies that

$$H_*(X_{hS^1}) \cong H_*\big((\Gamma v \otimes CU_*S^1, \tilde{\partial}) \underset{CU_*S^1}{\otimes} CU_*X\big),$$

we need to try to understand better the complex $(\Gamma v \otimes CU_*S^1, \tilde{\partial}) \underset{CU_*S^1}{\otimes} CU_*X$.

Define an extension $(\Gamma v \otimes CU_*X, D)$ of Γv by

$$D(v(n) \otimes U) := v(n) \otimes dU + \sum_{k=1}^{n-1}(-1)^k v(n-k-1) \otimes \kappa(T_k \otimes U).$$

We again use Proposition 3.1.6(2) to verify that $D^2 = 0$. Observe that $(\Gamma v \otimes CU_*X, D)$ is naturally a chain coalgebra that is a cofree left Γv-comodule, since κ is a coalgebra map and each T_k is primitive.

It is then easy to show that the following two maps are chain isomorphisms, one inverse to the other.

$$(\Gamma v \otimes CU_*S^1, \tilde{\partial}) \underset{CU_*S^1}{\otimes} CU_*X \longrightarrow (\Gamma v \otimes CU_*X, D)$$

$$v(n) \otimes U \otimes V \longmapsto v(n) \otimes \kappa(U \otimes V)$$

$$(\Gamma v \otimes CU_*X, D) \longrightarrow (\Gamma v \otimes CU_*S^1, \tilde{\partial}) \underset{CU_*S^1}{\otimes} CU_*X$$

$$v(n) \otimes V' \longmapsto v(n) \otimes 1 \otimes V'$$

Thus, $H_*(X_{hS^1}) \cong H_*(\Gamma v \otimes CU_*X, D)$.

In this chapter we are interested in cohomology calculations and so must dualize this model. Dualizing κ directly poses a problem, however, since

$$(CU_*S^1 \otimes CU_*X)^\sharp \ncong CU^*S^1 \otimes CU^*X$$

because the cubical chain complex on a space is not of finite type. We can avoid this problem by observing that it is enough to dualize the composition

$$< \mathcal{T} > \otimes CU_*X \underset{\simeq}{\overset{\iota}{\hookrightarrow}} CU_*S^1 \otimes CU_*X \xrightarrow{\kappa} CU_*X.$$

Let $j_n : \mathbb{Z} \cdot T_n \otimes CU_*X \longrightarrow < \mathcal{T} > \otimes CU_*X$ denote the natural inclusion of graded modules. Let $T_n^\sharp \in CU^{2n+1}S^1$ denote the cochain such that $T_n^\sharp(T_n) = 1$ and $T_n^\sharp(T) = 0$ if T is any other $(2n+1)$-cube. Let $< \mathcal{T} >^\sharp = \mathrm{Hom}(< \mathcal{T} >, \mathbb{Z})$

For each $n \geq 0$, define a linear map $\omega_n : CU^*X \longrightarrow CU^{*-(2n+1)}X$ of degree $-(2n+1)$ by

$$j_n^\sharp \circ (\kappa\iota)^\sharp(f) := T_n^\sharp \otimes \omega_n(f),$$

where

$$CU^*X \xrightarrow{(\kappa\iota)^\sharp} < \mathcal{T} >^\sharp \otimes CU^*X \xrightarrow{j_n^\sharp} \mathbb{Z} \cdot T_n^\sharp \otimes CU^*X.$$

In other words, ω_n is the dual of $\kappa(T_n \otimes -)$.

Let $(\Lambda v \otimes CU^*X, D^\sharp)$ denote the $\mathbb{Z}$-dual of $(\Gamma v \otimes CU_*X, D)$. In particular $v(v) = 1$. Since it is the dual of a cofree comodule, $(\Lambda v \otimes CU^*X, D^\sharp)$ is a free, right Λv-module. We need to identify $D^\sharp$ as precisely as possible, since

$$\mathrm{H}^*(\Lambda v \otimes CU^*X, D^\sharp) \cong \mathrm{H}^*(X_{hS^1}).$$

A simple dualization calculation gives us the following result.

Lemma 3.3.1. *If $f \in CU^m X$, then*

$$D^\sharp(\upsilon^n \otimes f) = \upsilon^n \otimes d^\sharp f + \sum_{k=0}^{\lceil \frac{m-1}{2} \rceil} (-1)^k \upsilon^{n+k+1} \otimes \omega_k(f)$$

where $d^\sharp$ denotes the differential of $CU^* X$.

As a consequence of this description of $D^\sharp$ we obtain the following useful properties of the operators ω_k.

Corollary 3.3.2. *The operators ω_n satisfy the following properties.*

(1) *For all $n \geq 1$, $[d^\sharp, \omega_n] = -\sum_{k=0}^{n-1} \omega_k \circ \omega_{n-k-1}$, while $d^\sharp \omega_0 = -\omega_0 d^\sharp$.*

(2) *Each ω_n is a derivation, i.e., $\omega_n(f \cdot g) = \omega_n(f) \cdot g + (-1)^f f \cdot \omega_n(g)$.*

Proof. The proof of (1) proceeds by expansion of the equation $0 = (D^\sharp)^2(1 \otimes f)$. To prove (2), expand the equation

$$D^\sharp(1 \otimes f \cdot g) = D^\sharp(1 \otimes f) \cdot (1 \otimes g) + (-1)^f (1 \otimes f) \cdot D^\sharp(1 \otimes g).$$

The differential $D^\sharp$ is a derivation, since it is the dual of the differential of a chain coalgebra. □

Remark 3.3.3. This corollary implies that ω_0 induces a derivation of degree -1

$$\varpi : \mathrm{H}^* X \longrightarrow \mathrm{H}^{*-1} X$$

such that $\varpi^2 = 0$.

3.4 The case of the free loop space

Let K be a finite-type, 1-reduced simplicial set such that $|K|$ has the same homotopy type as X. As we saw in section 2.4, there is commutative diagram

$$\begin{array}{ccc} C^*K & \xrightarrow{\iota} & C^*K\widetilde{\otimes}\mathcal{B}C^*K \\ \Big\downarrow \widetilde{\gamma}\sigma_K & & \simeq \Big\downarrow \Upsilon \\ CU^*X & \xrightarrow{CU^*e} & CU^*\mathcal{L}X \end{array}$$

in which $\widetilde{\gamma}\sigma_K$ is a quasi-algebra quasi-isomorphism, ι is a twisted bimodule extension and Υ is a quasi-isomorphism inducing an isomorphism on the E_∞-terms of the Eilenberg-Moore spectral sequence.

Our goal here is to combine this thin free loop space model with the general homotopy orbit space model of the section 3.3, obtaining an extension $(\Lambda\upsilon \otimes C^*K\widetilde{\otimes}\mathcal{B}C^*K, \widetilde{D})$ of $(\Lambda\upsilon, 0)$ by $fls^*(X)$, together with a quasi-isomorphism

$$\widetilde{\Upsilon} : (\Lambda\upsilon \otimes C^*K\widetilde{\otimes}\mathcal{B}C^*K, \widetilde{D}) \longrightarrow (\Lambda\upsilon \otimes CU^*\mathcal{L}X, D^\sharp)$$

such that

$$\begin{array}{ccccc} (\Lambda v, 0) & \xrightarrow{incl.} & (\Lambda v \otimes C^*K\widetilde{\otimes}\mathcal{B}C^*K, \widetilde{D}) & \xrightarrow{\pi} & C^*K\widetilde{\otimes}\mathcal{B}C^*K \\ \| & & \downarrow{\scriptstyle \tilde{\Upsilon}} & & \downarrow{\scriptstyle \Upsilon} \\ (\Lambda v, 0) & \xrightarrow{incl.} & (\Lambda v \otimes CU^*\mathcal{L}X, D^\sharp) & \xrightarrow{\pi} & CU^*\mathcal{L}X \end{array} \tag{3.1}$$

commutes, where π denotes the obvious projections.

We begin by an easy, though crucial, observation concerning the relations between Υ and the operations ω_k.

Lemma 3.4.1. *For all cocycles $x \in C^*K$, $\Upsilon(1 \otimes sx) = \omega_0 \Upsilon(x \otimes 1)$.*

Proof. From the definition of $\tilde{\beta}$ from Proposition 2.3.3 and of D from section 2.2, we can show that

$$\begin{aligned} \Upsilon(1 \otimes sx) =& CU^*j\tilde{\beta}(1 \otimes sx) \\ =& CU^*jJ\tilde{\beta}D(1 \otimes sx) \\ =& CU^*jJ\tilde{\beta}(s^{-1}(sx \otimes 1) - s^{-1}(1 \otimes sx)) \\ =& CU^*jJCU^*p\big(C^*pr_1 - C^*pr_2\big)\tilde{\gamma}(s^{-1}(sx)). \end{aligned}$$

A straightforward computation suffices to establish that

$$CU^*jJCU^*p\big(C^*pr_1 - C^*pr_2\big) = \omega_0 CU^*e,$$

implying that

$$\Upsilon(1 \otimes sx) = \omega_0 CU^*e\tilde{\gamma}(s^{-1}(sx)) = \omega_0 \Upsilon(x \otimes 1). \qquad \square$$

Restriction 3.4.2. Henceforth, to simplify the presentation, we assume that $\psi_K^\sharp$, the multiplication on $\mathcal{B}C^*K$, is such that the primitives of $\mathcal{B}C^*K$, i.e., the elements of sC^+K, are all indecomposable.

This is certainly a strong hypothesis, but it still allows us to treat a number of interesting cases, such as wedges of spheres. More general cases are treated in [H2].

The special properties of the free loop space model in the restricted case we consider are summarized in the following lemma.

Lemma 3.4.3. *If the primitives of $\mathcal{B}C^*K$ are all indecomposable, then the following properties hold.*

(1) *The multiplication on $\mathcal{B}C^*K$ is the shuffle product, which is commutative.*

(2) *The graded algebra C^*K is commutative.*

(3) *For all* $y, x_1, ..., x_n \in C^*K$,

$$\begin{aligned}\check{D}(y \otimes sx_1|\cdots|sx_n) &= dy \otimes sx_1|\cdots|sx_n + (-1)^y y \otimes d_{\mathcal{B}}(sx_1|\cdots|sx_n)\\ &\quad + (-1)^y\big[yx_1 \otimes sx_2|\cdots|sx_n - (-1)^N yx_n \otimes sx_1|\cdots|sx_{n-1}\big],\end{aligned}$$

where $N = (1+\deg x_n)(n-1+\sum_{1\le j<n}\deg x_j)$, *and*

$$(y\otimes 1)(1\otimes sx_1|\cdots|sx_n) = y \otimes sx_1|\cdots|sx_n.$$

In other words, the differential of $C^*K\widetilde{\otimes}\mathcal{B}C^*K$ is exactly that of the usual Hochschild complex on C^*K, while the left C^*K-action is untwisted, when the primitives of $\mathcal{B}C^*K$ are all indecomposable.

Proof. (1) This is obvious. (2) Recall from section 2.2 that if $\psi_K^\sharp$ is commutative, then

$$(1\otimes c)(1\otimes c') = 1 \otimes c\star c'.$$

Thus, if $x\in C^{m+1}K, y \in C^{n+1}K$, then

$$\begin{aligned}\check{D}(1\otimes sx\star sy) =&\check{D}(1\otimes sx)\cdot(1\otimes sy) + (-1)^m(1\otimes sx)\cdot\check{D}(1\otimes sy)\\ =&-(1\otimes s(dx))\cdot(1\otimes sy) - (-1)^m(1\otimes sx)\cdot(1\otimes s(dy)),\end{aligned}$$

whenever $\psi_K^\sharp$ is commutative. If, moreover, all primitives of $\mathcal{B}C^*K$ are indecomposable, then

$$\begin{aligned}\check{D}(1\otimes sx\star sy) =&\check{D}(1\otimes sx|sy + (-1)^{mn}1\otimes sy|sx)\\ =&x\otimes sy - (-1)^{mn}y\otimes sx + (-1)^m 1\otimes s(xy)\\ &-1\otimes s(dx)|sy - (-1)^m 1\otimes sx|s(dy)\\ &+(-1)^{mn}\big[y\otimes sx - (-1)^{mn}x\otimes sy + (-1)^n 1\otimes s(yx)\\ &\qquad -1\otimes s(dy)|sx - (-1)^n 1\otimes sy|s(dx)\big]\\ =&(-1)^m 1\otimes s([x,y]) - 1\otimes s(dx)\star sy - (-1)^m 1\otimes sx\star s(dy)\end{aligned}$$

and so $[x,y] = 0$. Hence, C^*K is commutative if all primitives of $\mathcal{B}C^*K$ are indecomposable.

(3) When all primitives of $\mathcal{B}C^*K$ are indecomposable, the formulas of section 2.4 obviously reduce to those given in the statement. □

We now define the desired extension

$$(\Lambda v, 0) \longrightarrow (\Lambda v\otimes C^*K\otimes \mathcal{B}C^*K, \widetilde{D}) \longrightarrow C^*K\widetilde{\otimes}\mathcal{B}C^*K$$

and show that

$$\mathrm{H}^*(\Lambda v\otimes C^*K\otimes\mathcal{B}C^*K,\widetilde{D}) \cong \mathrm{H}^*(X_{hS^1}).$$

We define the extension by

$$\widetilde{D} = Id \otimes \check{D} + v \cdot - \otimes S$$

where

(1) $S(y \otimes 1) = 1 \otimes sy$ for all $y \in C^+K$;

(2) $S(1 \otimes c) = 0$ for all $c \in \mathcal{B}C^*K$; and

(3) $S(y \otimes sx_1|\cdots|sx_n) = \sum_{j=1}^{n+1} \pm 1 \otimes sx_j|\cdots|sx_n|sy|sx_1|\cdot|sx_{j-1}$, where the sign is chosen in accord with the Koszul convention (cf., section 1.1).

Thus, for example, if $x, y, z \in C^*K$ are of degrees $l+1$, $m+1$ and $n+1$, respectively, then $S(x \otimes sy) = 1 \otimes (sx|sy + (-1)^{lm} sy|sx) = 1 \otimes sx \star sy$ and

$$S(x \otimes sy|sz) = 1 \otimes (sx|sy|sz + (-1)^{(m+l)n} sz|sx|sy + (-1)^{l(m+n)} sy|sz|sx).$$

It is obvious that $S^2 = 0$. A tedious, though not difficult, combinatorial calculation, shows that $\check{D}S = -S\check{D}$. The proof of this equality depends strongly on the fact that C^*K is commutative; in the general case we need to add terms to $S(x\otimes c)$ to kill certain commutators [H2]. Thus $\widetilde{D}^2 = 0$, i.e., $(\Lambda v \otimes C^*K \odot \mathcal{B}C^*K, \widetilde{D})$ is a cochain complex. Indeed this is exactly the negative cyclic complex of the commutative algebra C^*K, looked at as a cochain complex in positive degrees, rather than as a chain complex in negative degrees.

As Jones proved in [J], $\mathrm{H}^*(\mathcal{L}X_{hS^1})$ is isomorphic to the negative cyclic homology of the algebra S^*X, and therefore to that of C^*K, if $|K| \simeq X$. Thus

$$\mathrm{H}^*(\Lambda v \otimes C^*K \widetilde{\otimes} \mathcal{B}C^*K, \widetilde{D}) \cong \mathrm{H}^*(\Lambda v \otimes C^*\mathcal{L}X, D^\sharp) \cong \mathrm{H}^*(\mathcal{L}X_{hS^1}),$$

as desired. To build our model for topological cyclic homology, however, we need a cochain quasi-isomorphism $\widetilde{\Upsilon}$ lifting Υ and inducing this cohomology isomorphism. In the next theorem, we prove the existence of $\widetilde{\Upsilon}$ when K is an odd-dimensional sphere. Using results from [P], we can generalize this theorem to wedges of odd spheres and, when working over $\mathbb{F}_2$, to wedges of even spheres. The essential ideas of the general proof are already present in the proof for a single odd sphere, so we restrict to this case, to simplify the presentation. In [H2], we prove the existence of $\widetilde{\Upsilon}$ for a somewhat larger class of spaces.

Before stating and proving the theorem, we analyze carefully the S^1- action on $\mathcal{L}S^{2n+1}$. It is well known that $\mathrm{H}^*(\mathcal{L}S^{2n+1})$ is isomorphic to the tensor product of an exterior algebra Λx on a generator of degree $2n+1$ with the divided powers algebra Γy on a generator of degree $2n$ (cf., e.g., [Sm]). The generator x is represented by $CU^*e(\zeta)$, where $\zeta \in CU^{2n+1}S^{2n+1}$ represents the fundamental class of S^{2n+1}. More explicitly, if $U : I^{2n+1} \longrightarrow S^{2n+1}$ is a $(2n+1)$-cube collapsing ∂I^{2n+1} to a point, then

$$\zeta : CU_{2n+1}S^{2n+1} \longrightarrow \mathbb{Z}$$

is specified by $\zeta(U) = 1$ and $\zeta(V) = 0$ if V is any other $(2n+1)$-cube. Consider the transpose of U

$$U^\flat : I^{2n} \longrightarrow (S^{2n+1})^I : (t_1, ..., t_{2n}) \longmapsto U(-, t_1, ..., t_{2n}) .$$

Since U collapses the boundary of the cube, $U^\flat(t_1, ..., t_{2n})$ is always a (based) loop. Let $\xi \in CU^{2n}\mathcal{L}X$ be the cochain such that $\xi(U^\flat) = 1$ and $\xi(V) = 0$ for any other $2n$-cube V. The generator y is represented by ξ.

Recall the definition of ϖ at the end of section 3.3. A simple, explicit calculation shows that $\varpi(x) = y$, which implies that $\varpi(y) = 0$, whence

$$\varpi(x \otimes y(m)) = (m+1) \cdot y(m+1) \quad \text{and} \quad \varpi(1 \otimes y(m)) = 0$$

for all $m \geq 0$, since ϖ is a derivation. In particular, $\varpi(\mathrm{H}^{\text{even}}\, \mathcal{L}S^{2n+1}) = 0$, and $\varpi : \mathrm{H}^{\text{odd}}\, \mathcal{L}S^{2n+1} \longrightarrow \mathrm{H}^{\text{even}}\, \mathcal{L}S^{2n+1}$ is an isomorphism.

Theorem 3.4.4. *Let K be the simplicial model of S^{2n+1} with exactly two nondegenerate simplices, in degrees 0 and $2n+1$, where $n > 0$. There is a quasi-isomorphism $\widetilde{\Upsilon} : (\Lambda v \otimes C^*K \widetilde{\otimes} \mathcal{B}C^*K, \widetilde{D}) \longrightarrow (\Lambda v \otimes CU^*\mathcal{L}X, D^\sharp)$ such that*

$$\begin{array}{ccccc}
(\Lambda v, 0) & \xrightarrow{incl.} & (\Lambda v \otimes C^*K\widetilde{\otimes}\mathcal{B}C^*K, \widetilde{D}) & \xrightarrow{\pi} & C^*K\widetilde{\otimes}\mathcal{B}C^*K \\
\| & & \downarrow{\scriptstyle \widetilde{\Upsilon}} & & \downarrow{\scriptstyle \Upsilon} \\
(\Lambda v, 0) & \xrightarrow{incl.} & (\Lambda v \otimes CU^*\mathcal{L}S^{2n+1}, D^\sharp) & \xrightarrow{\pi} & CU^*\mathcal{L}S^{2n+1}
\end{array}$$

commutes, where π denotes the obvious projection maps.

Proof. In this case, $C^*K = \Lambda z$, an exterior algebra on an odd generator of degree $2n+1$ and $\mathcal{B}C^*K = \Gamma(sz)$, the divided powers algebra on an even generator of degree $2n$. Furthermore, $\check{D} = 0$, i.e., $(C^*K\widetilde{\otimes}\mathcal{B}C^*K, \check{D}) = (\Lambda z \otimes \Gamma sz, 0)$.

We need to define a cochain map $\widetilde{\Upsilon} = \sum_{k\geq 0} v^k \otimes \Upsilon_k$, where $\Upsilon_0 = \Upsilon$ and, for all k, $\Upsilon_k : C^*K\widetilde{\otimes}\mathcal{B}C^*K \longrightarrow CU^{*-2k}\mathcal{L}S^{2n+1}$ is a linear map of degree $-2k$. Since $\widetilde{D} = 1 \otimes \check{D} + v \otimes S$ and $D^\sharp = 1 \otimes d^\sharp + \sum_{k\geq 0} v^{k+1} \otimes \omega_k$, the equation $D^\sharp\widetilde{\Upsilon} = \widetilde{\Upsilon}\widetilde{D}$ is equivalent to the set of equations

$$d^\sharp \Upsilon_k + \sum_{i+j=k-1} \omega_i \Upsilon_j = \Upsilon_k \check{D} + \Upsilon_{k-1} S \tag{III.2$_k$}$$

for $k \geq 0$. Thus, to define $\widetilde{\Upsilon}$, we can build up the family of Υ_k's by induction on both k and wordlength in $\mathcal{B}C^*K = \Gamma sx$.

For $k = 0$, the equation above becomes simply $d^\sharp\Upsilon_0 = \Upsilon_0\check{D}$, which holds because Υ is a cochain map.

When $k = 1$, the appropriate equation is

$$d^\sharp \Upsilon_1 = \Upsilon_1 \check{D} + \Upsilon_0 S - \omega_0 \Upsilon_0 \qquad (\mathrm{III.2}_1)$$

Applied to $z \otimes 1 \in C^*K \widetilde{\otimes} \mathcal{B}C^*K$, the right-hand side of this equation becomes

$$0 + \Upsilon(1 \otimes sz) - \omega_0 \Upsilon(z \otimes 1),$$

which is 0, by Lemma 3.4.1. Thus, we can set

$$\Upsilon_1(z \otimes 1) = 0.$$

When we apply the right-hand side of $(\mathrm{III.2}_1)$ to $1 \otimes sz$, we obtain

$$0 + 0 - \omega_0 \Upsilon(1 \otimes sz),$$

which is equal to $-\omega_0^2 \Upsilon(z \otimes 1)$, by Lemma 3.4.1. We can therefore choose

$$\Upsilon_1(1 \otimes sz) = \omega_1 \Upsilon(z \otimes 1),$$

by Corollary 3.3.2 (1). By a similar argument, we can choose

$$\Upsilon_1(z \otimes sz) = \omega_1 \Upsilon(z \otimes 1) \cdot \Upsilon(z \otimes 1).$$

Suppose that Υ_1 has been defined on $\Lambda z \otimes \Gamma^{\leq l-1}(sz)$ satisfying equation $(\mathrm{III.2}_1)$, where $l \geq 2$. Applying the right-hand side of $(\mathrm{III.2}_1)$ to $1 \otimes sz(l)$, we obtain

$$0 + 0 - \omega_0 \Upsilon(1 \otimes sz(l)),$$

which is a cycle of odd degree. Since ϖ is injective on odd cohomology, either $\omega_0(1 \otimes sz(l))$ is a boundary or $\omega_0^2(1 \otimes sz(l))$ is not a boundary. The second option is impossible, since $\omega_0^2(1 \otimes sz(l)) = -d^\sharp \omega_1(1 \otimes sz(l))$, and so $\omega_0(1 \otimes sz(l))$ must be a boundary, i.e., there is possible choice of $\Upsilon_1(1 \otimes sz(l))$ satisfying $(\mathrm{III.2}_1)$. If we then set

$$\Upsilon_1(z \otimes sz(l)) = \Upsilon_1(1 \otimes sz(l)) \cdot \Upsilon(z \otimes 1),$$

then equation $(\mathrm{III.2}_1)$ is satisfied on $\Lambda z \otimes \Gamma^{\leq l}(sz)$.

Suppose now that for all $k < m$, there is a linear map

$$\Upsilon_k : \Lambda z \otimes \Gamma sz \longrightarrow CU^{*-2k} \mathcal{L} S^{2n+1}$$

satisfying $(\mathrm{III.2}_k)$ and such that

$$\Upsilon_k(z \otimes 1) = 0 \quad \text{and} \quad \Upsilon_k(1 \otimes sz) = \omega_k \Upsilon(z \otimes 1).$$

For $k = m$, the equation we must satisfy is

$$d^\sharp \Upsilon_m = \Upsilon_m \check{D} + \Upsilon_{m-1} S - \sum_{i+j=m-1} \omega_i \Upsilon_j. \qquad (\mathrm{III.2}_m)$$

Applied to $z \otimes 1$, the right-hand side of the equation becomes

$$0 + \Upsilon_{m-1}(1 \otimes sz) - \sum_{i+j=m-1} \omega_i \Upsilon_j(z \otimes 1),$$

which is zero, by the induction hypotheses. We can therefore set $\Upsilon_m(z \otimes 1) = 0$.

If the right-hand side of the equation (III.2_m) is evaluated on $1 \otimes sz$, it becomes

$$\begin{aligned} 0 + 0 - \sum_{i+j=m-1} \omega_i \Upsilon_j(1 \otimes sz) &= - \sum_{i+j=m-1} \omega_i \omega_j \Upsilon(z \otimes 1) \\ &= d^\sharp \omega_m \Upsilon(z \otimes 1), \end{aligned}$$

implying that we may set $\Upsilon(1 \otimes sz) = \omega_m \Upsilon(z \otimes 1)$.

Suppose that Υ_m has been defined on $\Lambda z \otimes \Gamma^{\leq l-1}(sz)$ satisfying equation (III.2_m), where $l \geq 2$. Applying the right-hand side of (III.2_m) to $1 \otimes sz(l)$, we obtain

$$0 + 0 - \sum_{i+j=m-1} \omega_i \Upsilon(1 \otimes sz(l)),$$

which is a cycle of odd degree. Since ϖ is injective on odd cohomology, either $\sum_{i+j=m-1} \omega_i \Upsilon_j(1 \otimes sx(l))$ is a boundary or its image under ω_0 is not a boundary. Observe however that

$$\begin{aligned} d^\sharp &\Big(\sum_{i+j=m-1} \omega_{i+1} \Upsilon_j(1 \otimes sz(l)) \Big) \\ &= \sum -\omega_s \omega_t \Upsilon_j(1 \otimes sz(l)) + \sum \omega_{i+1} \omega_p \Upsilon_q(1 \otimes sz(l)) \\ &= \sum_{i+j+k=m-1} -\omega_i \omega_j \Upsilon_k(1 \otimes sz(l)) + \sum \omega_i \omega_j \Upsilon_k(1 \otimes sz(l)) \\ &= - \sum_{i+j=m-1} \omega_0 \omega_i \Upsilon_j(1 \otimes sz(l)) \\ &= - \omega_0 \Big(\sum_{i+j=m-1} \omega_i \Upsilon_j(1 \otimes sz(l)) \Big), \end{aligned}$$

and so $\sum_{i+j=m-1} \omega_i \Upsilon_j(1 \otimes sz(l))$ must be a boundary. Hence, there is possible choice of $\Upsilon_m(1 \otimes sz(l))$ satisfying (III.2_m). If we then set

$$\Upsilon_m(x \otimes sz(l)) = \Upsilon_m(1 \otimes sz(l)) \cdot \Upsilon(z \otimes 1),$$

then equation (III.2_m) is satisfied on $\Lambda z \otimes \Gamma^{\leq l}(sz)$. □

Definition 3.4.5. Let X be a 1-connected space with the homotopy type of a finite-type CW-complex, and let $\Upsilon : C^* K \widetilde{\otimes} \mathcal{B}C^* K \xrightarrow{\simeq} CU^* \mathcal{L}X$ be a thin free loop model for X such that all primitives of $\mathcal{B}C^*K$ are indecomposable. The twisted bimodule extension

$$hos^*(X) = (\Lambda v \otimes C^*K \widetilde{\otimes} \mathcal{B}C^*K, \widetilde{D})$$

together with the quasi-isomorphism

$$\widetilde{\Upsilon} : hos^*(X) \longrightarrow (\Lambda v \otimes CU^*\mathcal{L}X, D^\sharp)$$

such that diagram 3.1 commutes, when it exists, is a *thin model* of $\mathcal{L}X_{hS^1}$.

Related work 3.4.6. Bökstedt and Ottosen have recently developed an approach to Borel cohomologly calculations for free loop spaces that is Eckmann-Hilton dual to the approach considered here and thus complementary to our methods [BO2]. They have constructed a Bousfield-type spectral sequence that converges to the cohomology of $(\mathcal{L}X)_{hS^1}$. While our model is easiest to deal with for spaces with few cells, the elementary cases for their model are Eilenberg-MacLane spaces.

Chapter 4

A model for mod 2 topological cyclic homology

We begin this section by supplying the final piece of the machine with which we build a model of $TC(X;2)$: a model of the pth-power map, for $p=2$. We then use the machine to obtain an explicit and precise description of $tc^*(X)$. To conclude we illustrate the power of both the $tc^*(X)$ and the $hos^*(X)$ models, by applying them to computing $\mathrm{H}^*(\mathcal{L}S^{2n+1}_{hS^1})$ and $\mathrm{H}^*(TC(S^{2n+1};2);\mathbb{F}_2)$.

4.1 The pth- power map

The pth-power map, λ^p, on a free loop space $\mathcal{L}X$ sends any loop to the loop that covers the same image p times, turning p times as fast, i.e., for all $\ell \in \mathcal{L}X$ and for all $z \in S^1$

$$\lambda^p(\ell)(z) := \ell(z^p),$$

where we see S^1 as the set of complex numbers of norm 1.

There is another useful way to define λ^p. Let $\mathcal{L}X^{(p)}$ denote the pullback of the iterated diagonal $\Delta^{(p-1)}: X \longrightarrow X^p$ and of $e^p : (\mathcal{L}X)^p \longrightarrow X^p$, i.e., the elements of $\mathcal{L}X^{(p)}$ are sequences of loops $(\ell_1, ..., \ell_p)$ such that $\ell_i(1) = \ell_j(1)$ for all i, j. Let $e^{(p)} : \mathcal{L}X^{(p)} \longrightarrow X$ denote the map sending a sequence of loops to their common basepoint.

The iterated diagonal map on $\mathcal{L}X$ corestricts to $\Delta^{(p-1)} : \mathcal{L}X \longrightarrow \mathcal{L}X^{(p)}$, while concatenation of loops defines a map $\mu^{(p-1)} : \mathcal{L}X^{(p)} \longrightarrow \mathcal{L}X$, restricting to the usual iterated multiplication on ΩX. It is clear that the pth-power map factors through $\mathcal{L}X^{(p)}$, as $\lambda^p = \mu^{(p-1)}\Delta^{(p-1)}$. Furthermore, the following diagram of fibrations commutes.

$$\begin{array}{ccccc}
\Omega X & \xrightarrow{\Delta^{(p-1)}} & (\Omega X)^p & \xrightarrow{\mu^{(p-1)}} & \Omega X \\
\downarrow{\scriptstyle i} & & \downarrow{\scriptstyle i} & & \downarrow{\scriptstyle i} \\
\mathcal{L}X & \xrightarrow{\Delta^{(p-1)}} & \mathcal{L}X^{(p)} & \xrightarrow{\mu^{(p-1)}} & \mathcal{L}X \\
\downarrow{\scriptstyle e} & & \downarrow{\scriptstyle e^{(p)}} & & \downarrow{\scriptstyle e} \\
X & \xrightarrow{=} & X & \xrightarrow{=} & X
\end{array}$$

Using techniques similar to those applied in chapter 3, it is possible to show that there is a twisted bimodule extension $C^*K\widetilde{\otimes}(\mathcal{B}C^*K)^{\otimes p}$ and a quasi-isomorphism

$$\Upsilon' : C^*K\widetilde{\otimes}(\mathcal{B}C^*K)^{\otimes p} \xrightarrow{\simeq} CU^*(\mathcal{L}X^{(p)}).$$

This construction can be performed with sufficient naturality to ensure that the diagram

$$\begin{array}{ccc}
C^*K\widetilde{\otimes}(\mathcal{B}C^*K)^{\otimes p} & \xrightarrow{Id\otimes(\psi_K^\sharp)^{(p-1)}} & C^*K\otimes\mathcal{B}C^*K \\
\downarrow{\scriptstyle \Upsilon'} & & \downarrow{\scriptstyle \Upsilon} \\
CU^*(\mathcal{L}X^{(p)}) & \xrightarrow{\Delta^{(p-1)}} & CU^*(\mathcal{L}X)
\end{array}$$

commutes.

To complete the construction of a model of the pth-power map, we need only to find a model of $\mu^{(p-1)}$. Modeling $\mu^{(p-1)}$ is very technical in the general case, however, requiring a fine analysis of the images of Υ and Υ'. For certain spaces, we can nevertheless show relatively easily that an acceptable model of $\mu^{(p-1)}$ is $Id\otimes\chi^{(p-1)}$, where χ denotes the usual coproduct on $\mathcal{B}C^*K$, an esthetically pleasing result.

We show below that $Id\otimes(\psi_K^\sharp)^{(p-1)}\chi^{(p-1)}$ is a model of λ^p, at least when K is a simplicial model of an odd sphere. To simplify calculations somewhat, we consider here only the case $p=2$; the case of arbitrary p, for a larger class of spaces, can be found in [HR].

We verify first that our candidate to be a model of λ^2 is in fact a cochain map.

Proposition 4.1.1. *If K be a finite-type, 1-reduced simplicial set such that $\psi_K^\sharp$ is commutative, then $Id\otimes(\psi_K^\sharp\chi) : C^*K\widetilde{\otimes}\mathcal{B}C^*K \longrightarrow C^*K\widetilde{\otimes}\mathcal{B}C^*K$ is a cochain map.*

Proof. We need first to show that

$$(Id\otimes(\psi_K^\sharp\chi))\check{D} = \check{D}(Id\otimes(\psi_K^\sharp\chi)).$$

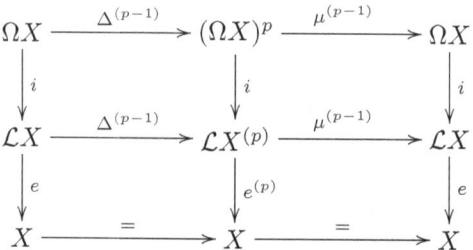

The computation, while combinatorially technical, is not subtle. The formulas in section 2.4 tell us that if $\psi_K^\sharp$ is commutative and $dx_i = 0$ for all i, then (up to signs)

$$\check{D}(1 \otimes sx_1|\cdots|sx_n) = \sum_{i=1}^{n-1}\Big[\pm\pi(sx_1 \star \cdots \star sx_{i-1} \star sx_n) \otimes sx_i|\cdots|sx_{n-1} \\ \pm \pi(sx_1 \star \cdots \star sx_i) \otimes sx_{i+1}|\cdots|sx_n \\ \pm 1 \otimes sx_1|\cdots|s(x_i x_{i+1})|\cdots|sx_n\Big]$$

and

$$(a \otimes 1)\cdot(1 \otimes sx_1|\cdots|sx_n) = a \otimes sx_1|\cdots|sx_n \\ + \sum_{i=1}^{n-1}\Big[\pm\pi(sa \star sx_1 \star \cdots \star sx_{i-1} \star sx_n) \otimes sx_i|\cdots|sx_{n-1} \\ \pm \pi(sa \star sx_1 \star \cdots \star sx_i) \otimes sx_{i+1}|\cdots|sx_n\Big],$$

while for all $a \in C^*K$

$$(Id \otimes (\psi_K^\sharp \chi))(a \otimes sx_1|\cdots|sx_n) = 2a \otimes sx_1|\cdots|sx_n \\ + a \otimes \sum_{i=1}^{n-1}(sx_1|\cdots|sx_i) \star (sx_{i+1}|\cdots|sx_n).$$

A bit of elementary algebra and careful counting enable us to show that $Id\otimes(\psi_K^\sharp\chi)$ is indeed differential, using the formulas above. □

Theorem 4.1.2. *Let K be the simplicial model of S^{2n+1} with exactly two nondegenerate simplices, in degrees 0 and $2n+1$, where $n > 0$. The diagram*

$$\begin{array}{ccc} C^*K\widetilde{\otimes}\mathcal{B}C^*K & \xrightarrow{Id\otimes(\psi_K^\sharp\chi)} & C^*K\widetilde{\otimes}\mathcal{B}C^*K \\ \simeq\downarrow\Upsilon & & \simeq\downarrow\Upsilon \\ CU^*\mathcal{L}S^{2n+1} & \xrightarrow{CU^*\lambda^2} & CU^*\mathcal{L}S^{2n+1} \end{array} \tag{4.1}$$

commutes up to cochain homotopy.

Proof. Recall that $\mathrm{H}^*(\mathcal{L}S^{2n+1}) = \Lambda x \otimes \Gamma y$, where $|x| = 2n+1$ and $|y| = 2n$. Furthermore $\mathrm{H}^*(\Omega S^{2n+1}) = \Gamma y$, and the restriction of λ^2 to ΩS^{2n+1} induces an endomorphism of Γy specified by

$$\begin{aligned}
H^*(\lambda^2|_{\Omega S^{2n+1}})(y(m)) &= H^*\chi H^*\mu(y(m))\\
&= H^*\chi(\sum_{i=0}^{m} y(i)\otimes y(m-i))\\
&= \sum_{i=0}^{m} y(i)\star y(m-i)\\
&= 2^m y(m).
\end{aligned}$$

Consequently, since $\lambda^2\circ i = i\circ\lambda^2|_{\Omega S^{2n+1}}$, the endomorphism $H^*\lambda^2$ of $\Lambda x\otimes\Gamma y$ induced by λ^2 must satisfy $H^*\lambda^2(1\otimes y(m)) = 2^m\otimes y(m)$, for degree reasons. Because $H^*\lambda^2$ is a map of algebras, it is therefore true that

$$H^*\lambda^2(x\otimes y(m)) = 2^m\cdot x\otimes y(m).$$

Let z denote the unique nondegenerate simplex of K in degree $2n+1$. As seen in section 3.4, the quasi-isomorphism

$$\Upsilon : C^*K\widetilde{\otimes}\mathcal{B}C^*K = (\Lambda z\otimes\Gamma sz, 0)\xrightarrow{\simeq} CU^*(\mathcal{L}S^{2n+1})$$

sends z to a representative ζ of x, sz to $\xi_1=\omega_0(\zeta)$, which represents y, and $sz(m)$ to some representative ξ_m of $y(m)$. Furthermore, calculations identical to those above show that

$$(Id\otimes\psi_K^\sharp\chi)(1\otimes sz(m)) = 2^m\otimes sz(m)\quad\text{and}\quad (Id\otimes\psi_K^\sharp\chi)(z\otimes sz(m)) = 2^m\cdot z\otimes sz(m).$$

Hence,

$$\Upsilon(Id\otimes\psi_K^\sharp\chi)(1\otimes sz(m)) = 2^m\Upsilon(1\otimes sz(m)) = 2^m\cdot\xi_m,$$

which is a representative of $2^m y(m)$, as is

$$CU^*\lambda^2\Upsilon(1\otimes sz(m)) = CU^*\lambda^2(\xi_m).$$

Since $y(m)$ is the unique class of degree $2nm$, there exists $\varsigma_m\in CU^{2nm-1}\mathcal{L}X$ such that

$$d^\sharp\varsigma_m = \Upsilon(Id\otimes\psi_K^\sharp\chi)(1\otimes sz(m)) - CU^*\lambda^2\Upsilon(1\otimes sz(m)),$$

which implies that

$$d^\sharp(\varsigma_m\cdot\zeta) = \Upsilon(Id\otimes\psi_K^\sharp\chi)(z\otimes sz(m)) - CU^*\lambda^2\Upsilon(z\otimes sz(m)).$$

Thus, the diagram 4.1 commutes up to a cochain homotopy G defined by $G(1\otimes sz(m)) = \varsigma_m$ and $G(z\otimes sz(m)) = \varsigma_m\cdot\zeta$. □

4.2 Topological cyclic homology

As explained in the Preface, we can now construct a cochain complex $tc^*(X)$ such that $\mathrm{H}^*(tc^*(X)\otimes\mathbb{F}_2)\cong \mathrm{H}^*((TC(X;2);\mathbb{F}_2)$, the mod 2 spectrum cohomology of $TC(X;2)$, at least for certain spaces X. The model $tc^*(X)$ is the mapping cone of the following composition, where π denotes the obvious projection map.

$$\begin{array}{ccc} (\Lambda v\otimes C^*K\widetilde{\otimes}\mathcal{B}C^*K,\widetilde{D}) & \xrightarrow{\pi} & C^*K\widetilde{\otimes}\mathcal{B}C^*K \\ & \searrow^{\tilde{\pi}} & \downarrow{\scriptstyle Id\otimes(Id-\psi^\sharp_K\chi)} \\ & & C^*K\widetilde{\otimes}\mathcal{B}C^*K \end{array}$$

Recall that the mapping cone of a cochain map $f:(V,d)\longrightarrow(V',d')$ is a cochain complex $C_f:=(V'\oplus sV,D_f)$, where $D_f(v')=d'v'$ for all $v'\in V'$ and $D_f(sv)=f(v)-s(dv)$ for all $sv\in sV$. It is an easy exercise to show that if f is cochain homotopic to g, then C_f and C_g are cochain equivalent. Theorem 4.1.2 suffices therefore to ensure that the mapping cone of the composition above has the right cohomology.

The next theorem now follows immediately from the results of the preceding chapters and section, according the justfication in [HR] of our method of construction of $tc^*(X)$.

Theorem 4.2.1. *Let X be a 1-connected space with the homotopy type of a finite-type CW-complex, and let $\Upsilon: C^*K\widetilde{\otimes}\mathcal{B}C^*K\xrightarrow{\simeq}CU^*\mathcal{L}X$ be a thin free loop model for X such that all primitives of $\mathcal{B}C^*K$ are indecomposable and such that*

$$\begin{array}{ccc} C^*K\widetilde{\otimes}\mathcal{B}C^*K & \xrightarrow{Id\otimes(\psi^\sharp_K\chi)} & C^*K\widetilde{\otimes}\mathcal{B}C^*K \\ \simeq\downarrow\Upsilon & & \simeq\downarrow\Upsilon \\ CU^*\mathcal{L}X & \xrightarrow{CU^*\lambda^2} & CU^*\mathcal{L}X \end{array}$$

commutes. Suppose that $\mathcal{L}X_{hS^1}$ has a thin model

$$\widetilde{\Upsilon}:(\Lambda v\otimes C^*K\widetilde{\otimes}\mathcal{B}C^*K,\widetilde{D})\xrightarrow{\simeq}(\Lambda v\otimes CU^*\mathcal{L}X,D^\sharp)\,.$$

Let

$$tc^*(X)=(C^*K\widetilde{\otimes}\mathcal{B}C^*K\oplus s(\Lambda v\otimes C^*K\widetilde{\otimes}\mathcal{B}C^*K),D_{\tilde{\pi}})$$

where for all $x\otimes c\in C^*K\otimes\mathcal{B}C^*K$,

$$D_{\tilde{\pi}}(x\otimes c)=\check{D}(x\otimes c)$$

while

$$\begin{aligned} D_{\tilde{\pi}}\big(s(1\otimes x\otimes c)\big)=&x\otimes c-x\otimes\psi^\sharp_K\chi(c)\\ &-s\big(1\otimes\check{D}(x\otimes c)+v\otimes S(x\otimes c)\big) \end{aligned}$$

and for $k>0$,

$$D_{\tilde{\pi}}\big(s(v^k\otimes x\otimes c)\big)=-s\big(v^k\otimes \check{D}(x\otimes c)+v^{k+1}\otimes S(x\otimes c)\big).$$

Then $\mathrm{H}^*(tc^*(X)\otimes \mathbb{F}_2)$ is isomorphic to the mod 2 spectrum cohomology of $TC(X;2)$.

A more general version of this theorem will appear in [HR].

We conclude this chapter and this article with an example illustrating the use of the models we have built.

Example 4.2.2. Let $n>0$, and let K be the model of S^{2n+1} with exactly one nondegenerate simplex z of positive dimension, in dimension $2n+1$. As explained in the proof of Theorem 3.4.4, $C^*K=\Lambda z$, with trivial differential. Furthermore, $\mathcal{B}C^*K$ is isomorphic as an algebra to Γsz, for degree reasons, and

$$fls^*(S^{2n+1})=(\Lambda z\otimes \Gamma sz,0).$$

Thus, $\mathrm{H}^*(\mathcal{L}S^{2n+1})\cong \Lambda z\otimes\Gamma sz$, as has long been known.

Let $sz(m)=sz|\cdots|sz\in\perp^m sz$. We then have

$$hos^*(S^{2n+1})=(\Lambda v\otimes\Lambda z\otimes\Gamma sz,\widetilde{D})$$

where $\widetilde{D}(v^k\otimes 1\otimes sz(m))=0$ for all k and m, while

$$\widetilde{D}(v^k\otimes z\otimes sz(m))=(m+1)v^{k+1}\otimes 1\otimes sz(m+1).$$

The integral cohomology of the homotopy orbit space is therefore

$$\mathrm{H}^*(\mathcal{L}S^{2n+1}_{hS^1})\cong\Lambda v\oplus\Gamma sz\oplus\bigoplus_{k,m\geq 1}\mathbb{Z}/m\mathbb{Z}\cdot(v^k\otimes sz(m))$$

as graded modules, while its mod p cohomology is of the form

$$\begin{aligned}\mathrm{H}^*(\mathcal{L}S^{2n+1}_{hS^1};\mathbb{F}_p)\cong&\Lambda v\otimes\mathbb{F}_p\oplus\Gamma sz\otimes\mathbb{F}_p\\&\oplus\bigoplus\mathbb{F}_p\cdot(v^k\otimes sz(m))\oplus\bigoplus\mathbb{F}_p\cdot(v^k\otimes z\otimes sz(m))\end{aligned}$$

where a Bockstein sends the class of $v^k\otimes z\otimes sz(m)$ to the class of $v^{k+1}\otimes sz(m+1)$.

Finally

$$tc^*(S^{2n+1})=(\Lambda z\otimes\Gamma sz\oplus s(\Lambda v\otimes\Lambda z\otimes\Gamma sz),D_{\tilde{\pi}})$$

where for all m,

$$D_{\tilde{\pi}}(z\otimes sz(m))=0=D_{\tilde{\pi}}(1\otimes sz(m))$$

while

$$\begin{aligned}D_{\tilde{\pi}}\big(s(1\otimes z\otimes sz(m))\big)=&z\otimes sz(m)-z\otimes\psi^\sharp_K\chi(sz(m))\\&-s\big(v\otimes S(z\otimes sz(m))\big)\\=&(1-2^m)z\otimes sz(m)-(m+1)s\big(v\otimes 1\otimes sz(m+1)\big)\end{aligned}$$

and

$$\begin{aligned}D_{\tilde{\pi}}\big(s(1 \otimes 1 \otimes sz(m))\big) &= 1 \otimes sz(m) - 1 \otimes \psi_K^\sharp \chi(sz(m)) \\ &= (1 - 2^m) \otimes sz(m)\end{aligned}$$

and for $k > 0$,

$$D_{\tilde{\pi}}\big(s(v^k \otimes z \otimes sz(m))\big) = -(m+1)s\big(v^{k+1} \otimes 1 \otimes sz(m+1))\big)$$

and

$$D_{\tilde{\pi}}\big(s(v^k \otimes 1 \otimes sz(m))\big) = 0.$$

Note that we use that

$$\psi_K^\sharp \chi(sz(m)) = \psi_K^\sharp(\sum_{i=0}^{m} sz(i) \otimes sz(m-i)) = \sum_{i=0}^{m} \binom{m}{i} sz(m) = 2^m sz(m).$$

Modulo 2, these formulas become

$$D_{\tilde{\pi}}(z \otimes sz(m)) = 0 = D_{\tilde{\pi}}(1 \otimes sz(m))$$

while

$$D_{\tilde{\pi}}\big(s(1 \otimes z \otimes sz(m))\big) = \begin{cases} z \otimes sz(m) & : m \text{ odd} \\ z \otimes sz(m) + s(v \otimes 1 \otimes sz(m+1)) \colon & m \text{ even} \end{cases}$$

and

$$D_{\tilde{\pi}}\big(s(1 \otimes 1 \otimes sz(m))\big) = 1 \otimes sz(m)$$

and for $k > 0$,

$$D_{\tilde{\pi}}\big(s(v^k \otimes z \otimes sz(m))\big) = \begin{cases} 0 & : m \text{ odd} \\ s(v^{k+1} \otimes 1 \otimes sz(m+1)) \colon & m \text{ even} \end{cases}$$

and

$$D_{\tilde{\pi}}\big(s(v^k \otimes 1 \otimes sz(m))\big) = 0.$$

We can now compute easily that

$$\begin{aligned}\mathrm{H}^*(TC(S^{2n+1};2);\mathbb{F}_2) \cong &\bigoplus \mathbb{F}_2 \cdot s(v^k \otimes sz(2m)) \oplus \mathbb{F}_2 \cdot s(v^k \otimes x \otimes sx(2m+1)) \\ &\oplus \bigoplus_{m \geq 0} \mathbb{F}_2 \cdot s(v \otimes sz(2m+1))\end{aligned}$$

as graded vector spaces.

Bibliography

[AH] J.F. Adams and P. J. Hilton, *On the chain algebra of a loop space* Comment. Math. Helv. **30**, (1956), 305–330.

[An] D. Anick, *Hopf algebras up to homotopy* J. Amer. Math. Soc. **2** (1989), 417–453.

[B] H.-J. Baues, *The cobar construction as a Hopf algebra* Invent. Math. **132** (1998), 467–489.

[Be] A.J. Berrick, *Algebraic K-theory and algebraic topology* Contemporary Developments in Algebraic K-Theory (ed. M Karoubi, A O Kuku, C Pedrini), ICTP Lecture Notes **15**, The Abdus Salam ICTP (Trieste, 2004), pp. 97–190.

[Bl] S. Blanc, *Modèles tordus d'espaces de lacets libres et fonctionnels* Thesis, EPFL (2004).

[BH] S. Blanc and K. Hess, *Simplicial and algebraic models for the free loop space* In preparation.

[BHM] M. Bökstedt, W.C. Hsiang, and I. Madsen, *The cyclotomic trace and algebraic K-theory of spaces* Invent. Math. **111** (1993) , 465–539.

[BO1] M. Bökstedt and I. Ottosen, *A splitting result for the free loop space of spheres and projective spaces* arXiv:math.AT/0411594

[BO2] M. Bökstedt and I. Ottosen, *A spectral sequence for string cohomology* arXiv:math.AT/0411571

[Br] E. H. Brown,*Twisted tensor products* Ann. Math. **69** (1959), 223–242.

[DH1] N. Dupont and K. Hess *Noncommutative algebraic models for fiber squares* Math. Annalen **314** (1999), 449–467.

[DH2] N. Dupont and K. Hess, *How to model the free loop space algebraically* Math. Annalen **314** (1999), 469–490.

[DH3] N. Dupont and K. Hess *Commutative free loop space models at large primes* Math. Z. **244** (2003), 1–34.

[DH4] N. Dupont and K. Hess *An algebraic model for homotopy fibers* Homology, Homotopy and Applications **4** (2002), 117–139.

[DS] W. G. Dwyer and J. Spaliński *Homotopy theories and model categories* Handbook of algebraic topology, North-Holland 1995, pp. 73–126.

[FHT] Y. Félix, S. Halperin, and J.-C. Thomas, *Rational Homotopy Theory* Graduate Texts in Mathematics **205**, Springer 2001.

[GJ] P. G. Goerss and J. F. Jardine, *Simplicial Homotopy Theory* Progress in Mathematics **174**, Birkhäuser 1999.

[GM] V.K.A.M Gugenheim, H. J. Munkholm, *On the extended functoriality of Tor and Cotor* J. Pure Appl. Algebra **4** (1974), 9–29.

[H] K. Hess,*Model categories in algebraic topology* Applied Categorical Structures **10** (2002), 195–220.

[H2] K. Hess, *Algebraic models of homotopy orbit spaces* In preparation.

[HPST] K. Hess, P.-E. Parent, J. Scott, and A. Tonks, *A canonical enriched Adams-Hilton model for simplicial sets* Submitted, 29 p.

[HR] K. Hess and J. Rognes, *Algebraic models for topological cyclic homology and Whitehead spectra of simply-connected spaces* In preparation.

[Ho] M. Hovey, *Model Categories* Mathematical Survey and Monographs **63**, American Mathematical Society, 1999.

[J] J.D.S. Jones, *Cyclic homology and equivariant homology* Invent. Math. **87** (1987), 403–423.

[K] K. Kuribayashi, *The cohomology of a pull-back on* $\mathbb{K}$*-formal spaces* Topology Appl. **125** (2002), 125–159.

[KY] K. Kuribayashi and T. Yamaguchi, *The cohomology algebra of certain free loop spaces* Fund. Math. **154** (1997), 57–73.

[Man] M. Mandell, E_∞*-algebras and* p*-adic homotopy theory* Topology **40** (2001), 43–94.

[Mas] W. Massey, *Singular Homology Theory* Graduate Texts in Mathematics **70**, Springer 1980.

[May] J. P. May, *Simplicial Objects in Algebraic Topology* University of Chicago Press 1967, Midway reprint 1982.

[Mc] J. McCleary, *A User's Guide to Spectral Sequences, Second Edition* Cambridge studies in advanced mathematics **58**, Cambridge University Press 2001.

[Me] L. Menichi, *On the cohomology algebra of a fiber* Algebr. Geom. Topol. **1** (2001), 719–742.

[Me2] L. Menichi, *The cohomology ring of free loop spaces* Homology Homotopy Appl.**3** (2001), 193–224.

[Mi] R. J. Milgram, *Iterated loop spaces* Ann. of Math. **84** (1966), 386–403.

[NT] B. Ndombol and J.-C. Thomas, *On the cohomology algebra of free loop spaces* Topology **41** (2002), 85–106.

[P] M. Parhizgar, *On the cohomology ring of the free loop space of a wedge of spheres* Math. Scand. **80** (1997), 195–248.

[R] J. Rognes, *The smooth Whitehead spectrum of a point at odd regular primes* Geom. Topol. **7** (2003), 55–184.

[Sm] L. Smith, *The Eilenberg-Moore spectral sequence and the mod* 2 *cohomology of certain free loop spaces* Ill. J. Math.**28** (1984), 516–522.

[SV] D. Sullivan and M. Vigué-Poirrier, *The homology theory of the closed geodesic problem* J. Diff. Geometry **11** (1976), 633–644.

[S] R. H. Szczarba, *The homology of twisted cartesian products* Trans. Amer. Math. Soc. **100** (1961), 197–216.

[W1] F. Waldhausen, *Algebraic K-theory of spaces* Algebraic and geometric topology (Proc. Conf., New Brunswick/USA 1983), Lect. Notes Math. **1126** (1985), pp. 318–419.

[W2] F. Waldhausen, B. Jahren, and J. Rognes, *The stable parametrized h-cobordism theorem* In preparation.